Stoke-on-Trent Libraries
Approved for Sale

STAFFORDSHIRE
COUNTY REFERENCE
LIBRARY
HANLEY
STOKE-ON-TRENT

Advanced Ceramics

Editor-in-Chief

Shinroku Saito
The Technical University of Nagaoka

Associate Editor

Kiyoshi Okazaki
The National Defence Academy

Editorial Committee (alphabetical order)

Hisao Banno
NGK Spark Plug Co., Ltd.

Anthony G. Evans
University of California

Charles D. Greskovich
General Electric Company

Shigeru Hayakawa
Matsushita Electric Ind. Co., Ltd.

Noboru Ichinose
Waseda University

Fumikazu Kanamaru
Osaka University

Mitsue Koizumi
Osaka University

Tomeji Ohno
NEC Corporation

Joseph A. Pask
University of California

Teiichi Yamazaki
TDK Corporation

Publishers' note
This work was originally published in 1987 by Ohmsha Ltd in Japan only under the title *Fine Ceramics*. To avoid confusion with other publishers' works this joint OUP/Ohmsha edition has been retitled. The content is unaltered.

Advanced Ceramics

Editor-in-Chief
SHINROKU SAITO

Copublished by
OXFORD UNIVERSITY PRESS
and
OHMSHA LTD
1988

Oxford University Press, Walton Street, Oxford OX2 6DP
Oxford New York Toronto
Delhi Bombay Calcutta Madras Karachi
Petaling Jaya Singapore Hong Kong Tokyo
Nairobi Dar es Salaam Cape Town
Melbourne Auckland
and associated companies in
Berlin Ibadan

Ohmsha Ltd, 3-1 Kanda Nishiki-cho, Chiyoda-ku, Tokyo 101, Japan

Oxford is a trademark of Oxford University Press

© *Ohmsha Ltd, 1988*

All rights reserved. No part of this publication may be reproduced,
stored in a retrieval system, or transmitted, in any form or by any means,
electronic, mechanical, photocopying, recording, or otherwise, without
the prior permission of Oxford University Press

British Library Cataloguing in Publication Data
Advanced ceramics.
1. Ceramics
I. Saito, Shinroku
666
ISBN 0-19-856335-3

Library of Congress Cataloging in Publication Data
Fine ceramics.
Advanced ceramics/editor-in-chief, Shinroku Saito.
p. cm.
Originally published: Fine ceramics. Tokyo: Ohmsha, 1987.
Includes bibliographies and index.
1. Ceramics. I. Saito, Shinroku, 1919–   . II. Title.
TP807.S24F56 1988 666-dc 19 88-21792
ISBN 0-19-856335-3

Printed in Great Britain
at the University Printing House, Oxford
by David Stanford
Printer to the University

Preface

Communication is the most important factor in creating deeper mutual understanding. However, the fundamental linguistic differences between Japanese and English often act as a barrier to improved communication. In light of the increasing importance of international cooperation, there have been many attempts to overcome this barrier.

The Ceramic Society of Japan has been requested to translate its journals into English concurrently with the publication of the monthly domestic edition. There are many proponents of such a translation not only abroad but also in Japan. One consideration is that society members deserve preferential treatment with respect to receiving the journal as soon as possible.

The journal of the Ceramic Society of Japan has been preparing numerical data, tables and abstracts written in English for more than 20 years. In addition, the Japan Fine Ceramics Association, which was established 5 years ago, began to publish a special annual report written in English. These actions continue to be useful in improving communications between speakers of English and Japanese. However, we editors feel that an additional step is required to further communications.

Therefore, we are presenting the first issue of "Fine Ceramics" in English. The editors invited papers not only from the limited circle of Japan but also from abroad. This conforms with our goal for improved international cooperation. The characteristic papers were also supplemented from Fulrath Awardees in Commemoration of the friendship of Late Prof. R. Fulrath of the University of California, Berkeley, U.S.A. In his regard a memorial comment is written herein by Prof. K. Okazaki of the Defence Academy of Japan.

The editor believes a brief comment on the term "Fine Ceramics" is necessary. The term "Fine" is apt to be taken as an adjective for art just as in fine art. In this sense, "Fine Ceramics" may be considered to be something such as beautiful fine porcelain. But in this case, the term "Fine" is intended to suggest a very minute grain structure composed of very minute and pure starting powder.

It was about 30 years ago when the first post war material boom took place internationally. Of course, Japan was involved in this boom. The main materials of that period were rare metals such as Te, Bi, etc. In the ceramic field, this is considered the first contemporary ceramic boom, mainly concerned with cermet. This boom, however, did not produce the anticipated results. There was another material boom approximately 10 years after the first one. The two booms, however, brought no promising results. The term "Fine", in the Japanese sense signifies the third boom, and anticipates a successful outcome.

Spring 1987

Shinroku SAITO
Editor-in-Chief

Contents

Preface

Introduction ······································· 1
    K. Okazaki; The National Defence Academy

Papers from Richard M. Fulrath Awardees

1. Recent Developments of Piezoelectric Composites in Japan ································· 8
   H. Banno; NGK Spark Plug Co., Ltd.
2. Research Trends in Ceramic Sensors ···················27
   N. Ichinose; Waseda University
3. Organic-Intercalation Compounds ····················41
   F. Kanamaru; Osaka University
4. Piezoelectric Ceramic Actuator ·······················62
   S. Takahashi; NEC Corporation

Original Review Papers

5. Toughening in ZrO_2-Based Materials ··················76
   K. T. Faber; The Ohio State University
6. A Review of Creep in Silicon Nitride and Silicon Carbide ·································95
   R. F. Davis and C. H. Carter, Jr.; North Carolina State University
7. The Glassy State ·································126
   J. F. Shackelford; University of California
8. Electrical Conductivity of Ceramic Materials ··········147
   D. M. Smyth; Lehigh University

9. Preparation ... n of Fine Powder
   from the Meta... ·················165
   Y. Ozaki; Seikei U...
10. Alumina Ceramic Substrates ·················184
    T. Ueyama and H. Wada;
    Hitachi Chemical Co., Ltd.
11. Bonding between Ceramics and Metals by HIPing······201
    M. Shimada; Tohoku University
    K. Suganuma, T. Okamoto and M. Koizumi;
    Osaka University
12. High Toughened PSZ (Partially Stabilized Zirconia) ··210
    T. Masaki and K. Kobayashi;
    Toray Industries, Inc.
13. Ceramic Application for Automotive Components ····227
    S. Wada; Toyota Central Research &
    Development Laboratories, Inc.
14. Pyroelectric Sensors of Lead Germanate Thick
    Films ·················240
    K. Takahashi; National Institute for Research
    in Inorganic Materials
15. Ferrite Materials·················254
    T. Nomura, K. Okutani and T. Ochiai;
    TDK Corporation

Authors' Profile ·················271

Keywords Index ·················277

Introduction

Kiyoshi OKAZAKI *

1. Purpose of This International Edition

In recent years, great scientific and technological advances in ceramics have been made, especially in the US and Japan, and the results of this research have been actively utilized in many different fields. Newly developed materials, fine ceramics in particular, are attracting greater attention for their superior characteristics. Applications in the mechanical, electrical, magnetic and electronics fields, as well as in biological and chemical fields are being extensively studied.

This volume on fine ceramics is edited by internationally renowned authorities in the field, and consists of papers which focus on activities in Japan as well as on research conducted in the US introducing the world's latest achievements in research and development of fine ceramics, from general fundamental methods to various applications.

There have been many fine research papers published in Japan, but, as they were published only in Japanese, valuable findings have often gone unnoticed in international circles. There were even unfortunate cases where research claims were subsequently reported as new findings by someone else in English. This English edition will provide a very significant publication opportunity for Japanese scholars.

2. Contents

Japan has done a significant amount of the original work in the field of Fine Ceramics (including electronic ceramics), two examples being the discovery of $BaTiO_3$ in 1944, and the development of Langevin-type $BaTiO_3$ ceramic vibrator for finding fish in 1951.[1] In 1950, Takagi proposed an antiferroelectric theory,[2] and after him, Sawaguchi et al.[3] conducted the first study of the phase diagram of $Pb(Zr, Ti)O_3$ (PZT). After B. Jaffe's patent was established there was a tremendous amount of research on piezoelectric compositions, including PZT, in Japan. Also, after the PSU (The Pennsylvania State University) group proposed "Coral structured PZT ceramics,"[4] many different types of composites have been studied in the US and Japan.

In Chapter 1, Banno reports on the recent development of piezoelectric composites and their applications for practical uses such as pressure sensors and hydrophone materials

---

* Vice President and Professor of Electronics Ceramics, The National Defense Academy, 1-10-20 Hashirimizu, Yokosuka, Kanagawa 239, Japan.

with a high voltage-output constant.

Takahashi developed a piezoelectric ceramic actuator with multilayer internal electrodes. The key point of this actuator is its uniform electrode configuration which eliminates stress concentration at the edges of the usual multilayer electrode structure. As a result, a maximum strain of about 0.1% can be detected. The actuators will be the next prospective field of applications of piezoelectric ceramics, as described in Chapter 4.

Ozaki describes the preparation and applications of fine powder from metal alkoxide for $BaTiO_3$ multilayer capacitors, $(Pb, La)(Zr, Ti)O_3$ (PLZT), and Al_2O_3 thin substrates in Chapter 9.

Sensors are also one of the important devices presented under study in the electronic industry. In Chapter 2, recent research trends in ceramic sensors in Japan are described by Ichinose. The NTC (negative temperature coefficient) thermistor is an old ceramic material; however, the practical application for clinical thermometer was started only a few years ago. Academic studies on PTC (positive temperature coefficient) of valency controlled $BaTiO_3$ semiconductive ceramics[5] have also been primarily studied in Japan since 1968. Both NTC and PTC ceramics are very important temperature sensors that use the temperature change of resistivity. In addition, this report describes pressure sensors utilizing various types of compositions and structures of piezoelectric ceramics, gas and humidity sensors that detect the absorption and deabsorption of gases by the changes in electrical conductivity, and infrared ray sensors utilizing the pyroelectric properties. These types of ceramic sensors promise a wider range of practical applications. Takahashi describes pyroelectric sensors of lead germanate thick films with the processing and the chemical and electric properties in Chapter 14.

Structural ceramics such as ZrO_2, SiC, Si_3N_4 and Al_2O_3 are attracting greater attention for their superior mechanical properties. The development of ceramic engine components is even being extensively studied. Evidence of this stems from the first Ceramic Engine Component Conference[6] held in 1983 in Hakone, organized by the Fine Ceramic Association.

In Chapter 5, a review of toughening in ZrO_2-based materials is described by Faber of the US. The main toughening mechanism comes from the martensitic transformation of tetragonal ZrO_2 particles to their monoclinic form. This mechanism is croucial in improving the mechanical behaviors of partially stabilized zirconia. In the editor's opinion, a part of this model will be useful for the understanding of mechanical properties of the ferroelectric ceramics with phase transition at the Curie point.

In Chapter 6, creep in silicone nitride and silicone carbide is reviewed by Davis et al. of the US. They describe the differences among the reaction-bonded, sintered and hot-pressed materials in the crystal structure, the microstructure, and the creep process.

In Chapter 7, Schakelford deals with the fundamentals of the glassy state such as optical properties, elastic behaviors with temperature, ionic transport and inert gas transport for microstructural understanding of fine ceramics.

Chapter 10 deals with alumina ceramic substrates. Ueyama describes a doctor blade method for their production. When compared to Chapters 4 and 5, this Chapter seems to be more technologically-oriented than the usual scientific paper. However, it should be emphasized that this kind of technology may prove to be very important for industrial production. This would, in turn, seem to indicate that technological originality should receive greater attention in the US. The point to be emphasized here is that one of the purposes of this series is to introduce technology originality in Japan. For practical production, manufacturing processes which deal with surface roughness, sintering and electrodes should ultimately converge to satisfy final applicaion requirements. In this regard, the "total technology" approach is very important.

Chapter 12 presents a Japanese review concerning partially stabilized zirconia, as described by Shimada et al.

In Chapter 13, Wada describes some ceramic applications for automotive electronic components and engine structures. The latter is discussed with regard to feasibility, reliability and cost, with emphasis upon the importance of post sinter machining. This Chapter includes some useful observations concerning the automobile industry of the 21st century.

In the famous book, entitled "Introduction to Ceramics," it is stated that in 1946, studies of J. L. Snoeck at the Philips Laboratories in Holland led to oxide ceramics with strong magnetic properties, high electrical resistivity, and low relaxation losses. Actually, Yogoro Kato (1882-1976), Emeritus Professor at Tokyo Institute of Technology, known as "the Father of Ferrite," and Takeshi Takei, Emeritus Professor at Keio University, also associated with initial ferrite research, first investigated[8] the chemical and magnetic properties of ferrites at the Tokyo Institute of Technology. Iron-cobalt ferrite-magnets and copper-zinc ferrite cores were subsequently produced by the Mitsubishi Electric Co., Ltd., in 1933, and by the TDK Electronic Co., Ltd. in 1935. Therefore, credit for this research actually belongs to Japan. Nevertheless, a major reason for such misunderstanding is language difficulties. As mentioned earlier, one of the purposes of this series is to minimize such barriers by introducing Japanese scientific and technological papers in English along with a 30% representation of American research papers.

Nomura et al. describes in Chapter 15, the recent development of ferrite materials with additional discussions of their production process, microstructure and magnetic properties.

In Chapter 8, Smyth reviews the electrical conductivity of ceramic materials in terms of their defect chemistry. Such a study is especially important for Japanese readers in order to understand the conduction mechanisms of such electronic ceramics as $BaTiO_3$ and TiO_2. The influences of nonstoichiometry and impurity additions sometimes can lead to a drastic improvement in characteristics. Examples of this include PTC materials and the non-reduced dielectric ceramic composition in a $N_2 + H_2$ atmosphere.[9] This material can be used for multilayer capacitors with base metal electrodes.

In Chapter 3, Kanamaru reports on organic intercalation compounds using organic substances with inorganic layered solid surfaces. This is one of the fundamental research areas in Japan at present.

One general impression gained from comparing the American and Japanese papers centers on the theoretical orientation of the papers from the US, while the Japanese papers are more technologically oriented. One of the reasons for this difference seems to originate in the basic difference in approach between the two research systems, as shown in Fig. 1.[10]

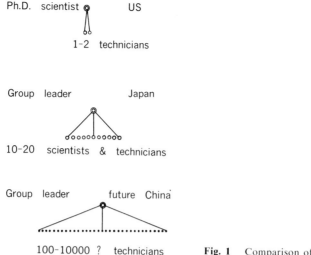

Fig. 1 Comparison of research project.

In the US system, a typical research project is conducted by a highly qualified Ph.D. scientist with the help of only 1 or 2 technicians, whereas a typical research project in Japan is carried out by a group of 10 to 20 competent reseachers and engineers headed by a group leader. The technological developments produced by the latter may, at times, appear not so scientifically valuable. Nevertheless, in commercial competition, technological originality must be considered as the most decisive and critical factor.

3. The Richard M. Fulrath Award Program

Several contributions have been made by Richard M. Fulrath Awardees. Professor Fulrath, formerly with University of California, Berkeley made a great contribution to the promotion of friendly relationship between the US and Japan through his work and display of humanity to Japanese scientists. After his death in July 1977, the Fulrath Memorial Meeting chaired by T. Yamazaki (President Emeritus, TDK Electronic Co., Ltd.) with S. Saito (President Emeritus, Tokyo Institute of Technology), M. Koizumi (Professor, Osaka University), S. Hayakawa (Matsushita Materials Research Laboratory) and K. Okazaki, established a Fulrath Memorial Fund. The fund was donated to the Department of Material Science and Minerals Engineering, University of California, Berkeley. The Richard M. Fulrath Award Committee was organized in the US by Joseph A. Pask, Chairman of the department, and assisted by his department's professors, the chairman and the vice-chairman of the northern California section of the American Ceramic Society and the President and Executive Director of the Society.

One scientist from Japanese academia, two from Japanese industries and one American scientist are selected each year as Fulrath awardees. The final selection of the awardees is made by the Berkeley committee from candidates under 45 years of age. The awardees of 1978-1985 are listed in Table 1.

The Fulrath Awards are presented on the Berkeley campus every October with the Fulrath Memorial Symposium held as an event of the Northern California Section of the

Table 1 The Richard M. Fulrath Awardees.

1978

F. Kanamaru; "Synthesis and Properties of Layered Organic-Inorganic Complexes"
  Professor, Institute of Scientific and Industrial Research, Osaka University, Suita, Osaka 565
N. Ichinose; "Development of Ceramic Gas Sensors"
  Department of Metal Engineering, Wasede University, Okubo, Shinjuku-ku, Tokyo 160
T. Ohno; "Properties of Ternary and Quaternary Systems Containing Pb(Zr, Ti)O_3"
  NEC Central Laboratory, Miyazaki, Miyamae-ku, Kawasaki, Kanagawa 213

1979

M. Inagaki; "Crystallization of Amorphous Carbon"
  Professor, Toyohashi University of Technology, Toyohashi, Aichi 440
Y. Imaoka; "Synthesis and Development of Magnetic Fine Powder"
  TDK Electronics Co., Ltd., 1-13-1 Nihonbashi, Chuo-ku, Tokyo 103
I. Ueda; "Development of PbTiO_3 Piezoelectric Ceramics
  Matsushita Materials Research Laboratory, Kadoma, Osaka 571
Anthony G. Evans; "Relationship between Ceramic Microstructures and Mechanical Properties"
  University of California, Santa Barbara, CA 94720

1980

T. Hirai; "Chemical Vapor Deposition of Non-Oxide Ceramics, especially for Silicone Nitride"
  Professor, The Research Institute for Iron, Steel and Other Metals, Tohoku University, Katahira, Sendai, Miyagi 980
H. Banno; "Development of Modified Pb(Zr, Ti)O_3 Ceramics"
  NGK Spark Plug Co., Ltd., Takatsuji-cho, Mizuho-ku, Nagoya, Aichi 467
H. Hirano; "Development of LiTaO_3 Crystal for Surface Acoustic Wave Devices"
  Toshiba Research & Development Center, Saiwai-ku, Kawasaki, Kanagawa 210
Richard C. Bradt; "Some Structure, Toughness and Elastic Modulus Trends in Glasses"
  University of Washington, Seattle, WA 98145

1981

T. Shiosaki; "Fabrication and Properties of Piezoelectric ZnO and AlN Thin Films"
  Associate Professor, Department of Electronics, Kyoto University, Sakyo-ku, Kyoto 606
M. Matsuoka; "Research and Development of Zinc Oxide Varistors"
  Matsushita Wireless Research Laboratory, Kadoma, Osaka 571
K. Tada; "Development and Applications of Electro-Optic Bismuth Silicone Oxide Single Crystal"
  Sumitomo Electric Industries, Ltd., 5-15 Kitahama, Higashi-ku, Osaka 541
H. Kent Bowen; "Pre-Sintering Science, Microengineering of Particulates"
  Massachusetts Institute of Technology, Cambridge, Boston, MA 02139

1982

A. Sawaoka; "Applications of High Pressure Technology for Very Hard Materials"
  Professor, Research Laboratory of Engineering Materials,
  Tokyo Institute of Technology, Nagatsuda, Midori-ku, Yokohama, Kanagawa 227
M. Furukawa; "Hot-Isostatic Pressing and Sintering Technology for Al$_2$O$_3$ Ceramic Tools"
  Nippon Tungsten Co., Ltd., 460 Shiobara-Sanno, Minami-ku, Fukuoka, Fukuoka 815
M. Yonezawa; "Development of Multilayer Ceramic Capacitor Materials"
  NEC Central Research Laboratory, Miyazaki, Miyamae-ku, Kawasaki, Kanagawa 213
F. F. Lange; "Fracture Mechanics in Si$_3$N$_4$ and SiC Ceramics"
  Rockwell International Science Center, Thousand Oaks, CA 91360

1983

K. Niihara; "Mechanical Properties of CVD Nonoxide Ceramics"
  The National Defence Academy, Yokosuka, Kanagawa 239
K. Komeya; "Developmant of Nitrogen Ceramics"
  Toshiba Research & Development Center, Saiwai-ku, Kawasaki, Kanagawa 210
Y. Kuwano; "Development of Integrated Amorphous-Silicon Solar Cells"
  Sanyo Central Lab., Hirakata, Osaka, Osaka 573
C. D. Greskovich; "Thermomechanical Properties of a New Composition of Sintered Si$_3$N$_4$"
  Corporate R & D, G. E., Schenectady, NY 12301

1984

S. Horiuchi; "Application of Atomic Resolution Electron Microscopy to Ceramic Materials"
  National Institute for Research in Inorganic Materials, Ibaraki 305
S. Yamazaki; "Metal-Insulator Semiconductor Structure"
  Semiconductor Energy Laboratory Co., Ltd., Atsugi, Kanagawa 243

H. Abe; "Mechanical Properties of Engineering Ceramics"
Asahi Glass Co., Ltd., Yokohama, Kanagawa 221

D. W. Johnson, Jr.; "Sol-Gel Processing of Ceramics and Glasses"
AT & T Bell Labs., Murray Hill, NJ 07974

1985

M. Shimada; "Structural Non-Oxide Ceramics by Hot-Pressing"
Tohoku University, Sendai, Miyagi 980

N. Yamaoka; "Strontium Titanate Based Boundary Layer Capacitors"
Taiyo Yuden Co., Ltd., Taito-ku, Tokyo 110

S. Takahashi; "Multilayer Piezoelectric Actuators and Their Applications"
NEC Research Lab., Miyazaki, Miyamae-ku, Kawasaki, Kanagawa 213

L. C. De Johghe; "Beta-Almina Solid Electrolytes"
Lawrence Berkeley Laboratory, University of California, Berkeley, CA94720

American Ceramic Society. The Fulrath Memorial Symposium is again repeated at the Pacific Coast Regional Meeting of the American Ceramic Society. The American awardee with the aid of the cash award visits Japan the following January for two to three weeks for scientific exchange and lectures at several universities, laboratories and industries. The purpose of this program is to stimulate friendly exchanges between the US and Japan on fundamental materials in ceramic science and technology.

References

1) K. Abe, T. Tanaka, I. Saito and K. Okazaki: Denki-Hyoron, **39**, 8 (1951) 12 [in Japanese].
2) Y. Takagi: Phys. Rev., **85** (1952) 315.
3) E. Sawaguchi: J. Phys. Soc. Japan, **8** (1953) 615.
4) D. P. Skinner, R. E. Newnham and L. E. Cross: Mater. Res. Bull., **13** (1978) 599.
5) O. Saburi: J. Am. Ceram. Soc., **44**, 2 (1961) 54.
6) S. Somiya, E. Kanai and K. Ando: Proc. 1st Int. Symp. on Ceramic Components for Engine (1983).
7) W. D. Kingery: Introduction to Ceramics, John Wiley & Sons, Inc., New York (1960).
8) Y. Kato and T. Takei: J. of Mining Soc. Japan, **46** (1930) 107.
9) Y. Sakabe, K. Minai and K. Wakino: Jpn. J. Appl. Phys., **20**, Suppl. 20-4 (1981) 147.
10) K. Okazaki: Japan-US Study Seminar on Dielectric and Piezoelectric Ceramics (1982) 0-1.

Papers from Richard M. Fulrath Awardees

1. Recent Developments of Piezoelectric Composites in Japan

Hisao BANNO*

Abstract

This paper describes developments of piezoelectric composite materials, their applications in Japan, and theoretical considerations for the dielectric and piezoelectric constants of composite types 1-3, 0-3, 2-2, and 3-0.

There is a newly derived theoretical equation for the piezoelectric d_{31} constant of the 1-3 type parallel model, which is distinguished from that of the conventional 2-2 type parallel one.

Ellipsoidal ceramic particle arrangement perpendicular to the plane surface in PVDF-PZT composites, which has been understood by Yamada et al. based on the ellipsoidal model, is discussed in view of the modified cubes model.

The dependencies of the dielectric and piezoelectric constants on the shape and volume fraction of pores are also discussed based on the modified cubes model.

Keywords: piezoelectric, composites, material, theory, applications.

1.1 Introduction

Early attempts to theoretically predict the dependencies of the electrical properties of the composites on the ceramic volume percent were made by Maxwell and Wagner[1] in 1914.

Such a theory has been derived from a spherical model.[1] Based on parallel and series models, theoretical equations for the dielectric constant have also been derived.[2]

In recent years piezoelectric composites have attracted much interest because of the performance limitations of single-phase materials.

Early attempts to make piezoelectric composites of ceramic particles and polymers were made in Japan by Kitayama et al.[3] and Pauer[4] in the United States.

In 1973 Pauer discussed the dielectric properties of the piezoelectric composites of urethane rubber and PZT ceramic particles based on series, parallel and spherical models, as well as his newly proposed cubes model. In 1976 a theoretical equation for the piezoelectric d_{31} constant was derived from a spherical model by Furukawa et al.[5]

---

* NTK Technical Ceramics Division, NGK Spark Plug Co., Ltd., 14-18, Takatsuji-cho, Mizuho-ku, Nagoya 467, Japan.

A concept of connectivity was introduced by Newnham et al. in 1978 and the composites have been classified according to this concept.[6] They also derived theoretical equations for the piezoelectric constants of the composites from series and parallel models.

This paper describes recent developments of piezoelectric composites in Japan in view of their theory and applications.

1.2 Theory

Theories have been developed based on some proposed models for composites, which are shown for a two phase system in Fig. 1.1.

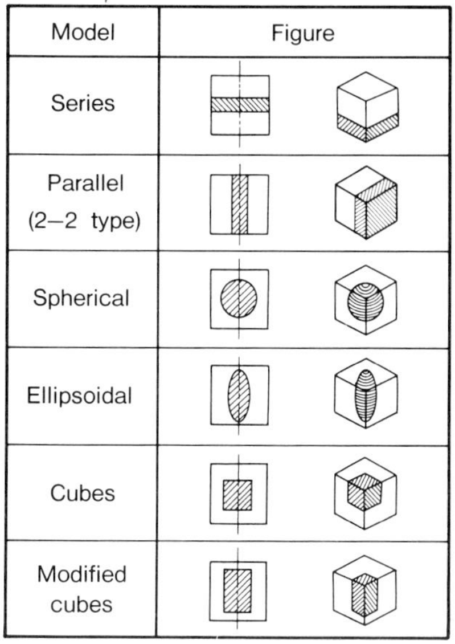

Model	Figure	
Series		
Parallel (2–2 type)		
Spherical		
Ellipsoidal		
Cubes		
Modified cubes		

Fig. 1.1 Schematic representation of proposed models for a two phase system of composites.

The spherical model for a two phase system is composed of a continuous matrix with spherical inclusions. Assuming that the shapes of the inclusions are cubic or ellipsoidal, the model becomes a cubes or ellipsoidal model.

Many theoretical works on the expression for the dielectric constant have been carried out, which are summarized in Table 1.1. The notation used in this paper is defined in Table 1.2.

Upper and lower limits of a two phase system have been obtained from equations based on the parallel and series models, respectively. According to Buessem,[2] general formulae for the intermediate state of the system are expressed as follows;

$$(\bar{\varepsilon}_{33})^r = {}^1v \cdot ({}^1\varepsilon_{33})^r + {}^2v \cdot ({}^2\varepsilon_{33})^r \tag{1.1}$$

where $-1 < r < +1$.

When r is nearly zero, $(\varepsilon_{33})^r$ becomes approximately $1 + r \cdot \log\varepsilon_{33}$. Accordingly, a "logarithmic mixture rule" appears as expressed in the following equation:

$$\log\bar{\varepsilon}_{33} = {}^1v \cdot \log{}^1\varepsilon_{33} + {}^2v \cdot \log{}^2\varepsilon_{33}. \tag{1.2}$$

Table 1.1 Theoretical equations for the dielectric constant of a two phase system (composite) derived from proposed models.

Model	Dielectric constant of composite	
Series	$1/\bar{\varepsilon}_{33} = {}^1v/{}^1\varepsilon_{33} + {}^2v/{}^2\varepsilon_{33}$	(T1.1)
Parallel	$\bar{\varepsilon}_{33} = {}^1v \cdot {}^1\varepsilon_{33} + {}^2v \cdot {}^2\varepsilon_{33}$	(T1.2)
Spherical	$\bar{\varepsilon}_{33} = {}^2\varepsilon_{33} \left\{ \dfrac{2 \cdot {}^2\varepsilon_{33} + {}^1\varepsilon_{33} - 2 \cdot {}^1v({}^2\varepsilon_{33} - {}^1\varepsilon_{33})}{2 \cdot {}^2\varepsilon_{33} + {}^1\varepsilon_{33} + {}^1v({}^2\varepsilon_{33} - {}^1\varepsilon_{33})} \right\}$	(T1.3)
Ellipsoidal	$\bar{\varepsilon}_{33} = {}^2\varepsilon_{33} \left\{ 1 + \dfrac{K \cdot {}^1v({}^1\varepsilon_{33} - {}^2\varepsilon_{33})}{K \cdot {}^2\varepsilon_{33} + ({}^1\varepsilon_{33} - {}^2\varepsilon_{33})(1 - {}^1v)} \right\}$ When the particles are spherical, $K=3$.	(T1.4)
Cubes	$\bar{\varepsilon}_{33} = \dfrac{{}^1\varepsilon_{33} \cdot {}^2\varepsilon_{33}}{({}^2\varepsilon_{33} - {}^1\varepsilon_{33}) \cdot {}^1v^{-1/3} + {}^1\varepsilon_{33} \cdot {}^1v^{-2/3}} + {}^2\varepsilon_{33} \cdot (1 - {}^1v^{2/3})$	(T1.5)
Modified Cubes	$\bar{\varepsilon}_{33} = \dfrac{a^2 \cdot [a + (1-a)n]^2 \cdot {}^1\varepsilon_{33} \cdot {}^2\varepsilon_{33}}{a \cdot {}^2\varepsilon_{33} + (1-a)n \cdot {}^1\varepsilon_{33}} + \{1 - a^2 \cdot [a + (1-a)n]\} \cdot {}^2\varepsilon_{33}$ where ${}^1v = a^3$. When $n = 0, 1$ and $(1/a^2 - a)/(1-a)$, it becomes the parallel, cubes and series model, respectively.	(T1.6)

Table 1.2 Notation used in this paper.

	Phase 1	Phase 2	Composite
Dielectric Constant	${}^1\varepsilon_{33}$ ${}^1\varepsilon_{22}$ ${}^1\varepsilon_{11}$	${}^2\varepsilon_{33}$ ${}^2\varepsilon_{22}$ ${}^2\varepsilon_{11}$	$\bar{\varepsilon}_{33}$ $\bar{\varepsilon}_{22}$ $\bar{\varepsilon}_{11}$
Piezoelectric d Constant	${}^1d_{33}$ ${}^1d_{32}$ ${}^1d_{31}$ ${}^1d_h$	${}^2d_{33}$ ${}^2d_{32}$ ${}^2d_{31}$ ${}^2d_h$	$\bar{d}_{33}$ $\bar{d}_{32}$ $\bar{d}_{31}$ $\bar{d}_h$
Elastic S Constant	${}^1S_{33}$ ${}^1S_{22}$ ${}^1S_{11}$	${}^2S_{33}$ ${}^2S_{22}$ ${}^2S_{11}$	$\bar{S}_{33}$ $\bar{S}_{22}$ $\bar{S}_{11}$
Mechanical Quality Factor	${}^1Q_{m33}$ ${}^1Q_{m11}$ ${}^1Q_{mp}$	${}^2Q_{m33}$ ${}^2Q_{m11}$ ${}^2Q_{mp}$	$\bar{Q}_{m33}$ $\bar{Q}_{m11}$ $\bar{Q}_{mp}$
Mechanical Loss Tangent	${}^1\tan\delta_{m33}$ ${}^1\tan\delta_{m11}$	${}^2\tan\delta_{m33}$ ${}^2\tan\delta_{m11}$	$\overline{\tan\delta_{m33}}$ $\overline{\tan\delta_{m11}}$
Frequency Constant	${}^1f_r \cdot d$	${}^2f_r \cdot d$	$\bar{f}_r \cdot d$
Radial Coupling Factor	${}^1k_p$	${}^2k_p$	$\bar{k}_p$
Volume Fraction	${}^1v$	${}^2v$	$\bar{v} = {}^1v + {}^2v = 1$

Although Eq.(1. 1) generalizes the mathematical expression for the parallel and series models, the author does not think that the r value has any physical meaning.

Pauer developed a theoretical expression for the dielectric constant based on the cubes model and he applied his theory to composites consisting of PZT ceramic particles and urethane rubber. The results are shown in Fig. 1. 2, where his experimental values did not agree with the calculated ones based on the parallel, series, cubes, and spherical models and the logarithmic mixture rule. The measured values are higher than the calculated ones based on the cubes and spherical models.

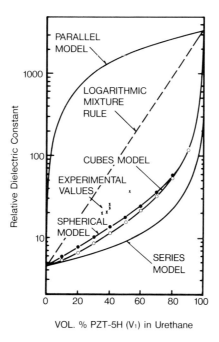

Fig. 1. 2 Comparison between experimental and theoretical values for dielectric constants of composites consisting of PZT-5H ceramic particles and urethane rubber. Theoretical values based on parallel, series, cubes and spherical models and on logarithmic mixture rule.

Banno (author of present paper) et al. has introduced a modified cubes model,[7] which generalizes the parallel, series and cubes models as illustrated in Fig. 1. 3. In Fig. 1. 3, l, m, n are parameters attributed to the shape of the unit cell deformed from the cubes one.

In many cases, l is equal to m. In this case, the modified cubes model is determined by the values of n and a, and the volume fraction of phase 1 (^1v) is a^3. When $n = 0$, 1 and $(^1v^{-2/3} - {}^1v^{1/3})/(1 - {}^1v^{1/3})$ ($l = m = 0$), the modified cubes model becomes the parallel, cubes and series ones respectively as shown in the Fig. 1. 3.

That phase 1 is always cubic and therefore the unit cell cannot always be cubic is shown in Fig. 1. 3(A). If this unit cell is equivalently transformed to be cubic, then phase 1 will not always be cubic as shown in Fig. 1. 3(B).

The theoretical equations for the piezoelectric constants of the composites were first derived from the spherical model by Furukawa et al.[5] in 1976. They were developed in 1982

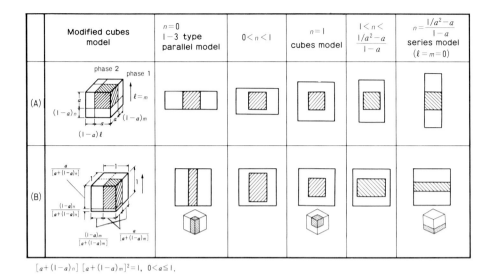

Fig. 1. 3 Schematic representation of modified cubes model as a function of the n value when (A) phase 1 is cubic and (B) the unit cell is cubic.

based on the ellipsoidal model by Yamada et al.,[8] who applied their theory to the piezoelectric composite material of PZT ceramic particles and PVDF polymers. The theoretical equations derived from the above-mentioned two models are shown in Table 1. 3, where K is a parameter attributed to the shape of the ellipsoidal particles.

Table 1. 3 Theoretical equations for the piezoelectric d_{31} constant of a two phase system (composite), derived from spherical and ellipsoidal models, when $^2d_{31}=0$, $^2S_{11} \gg {}^1S_{11}$ and $^2S_{33} \gg {}^1S_{33}$.

Model	Piezoelectric d_{31} constant of composite	
Spherical	$\bar{d}_{31} \cong {}^1d_{31} \cdot \dfrac{5 \cdot {}^1v}{2 + 3 \cdot {}^1v} \cdot \dfrac{3 \cdot {}^2\varepsilon_{33}}{2 \cdot {}^2\varepsilon_{33} + {}^1\varepsilon_{33} + {}^1v({}^2\varepsilon_{33} - {}^1\varepsilon_{33})}$	(T3.1)
Ellipsoidal	$\bar{d}_{31} = \dfrac{\alpha \cdot {}^1v \cdot K \{{}^1\varepsilon_{33}/{}^2\varepsilon_{33} - 1 + K + (K-1)({}^1\varepsilon_{33}/{}^2\varepsilon_{33} - 1) \cdot {}^1v\} \cdot {}^1d_{31}}{({}^1\varepsilon_{33}/{}^2\varepsilon_{33} - 1 + K)^2 + ({}^1\varepsilon_{33}/{}^2\varepsilon_{33} - 1)\{(K-1)^2 - {}^1\varepsilon_{33}/{}^2\varepsilon_{33}\} \cdot {}^1v}$	(T3.2)
	where α is polarizing factor. When the particles are spherical, $K=3$.	

They reported that the theoretical values agreed well with the measured ones in both dielectric and piezoelectric experiments when the parameter K was 8.5. The results, wherein the parameter K was 8.5, demanded that the ellipsoidal axes ratio was 2.8 and that the long ellipsoids of PZT powder were perpendicular to the composite film surface.[8]

However, it is difficult to accept these specifications because it is most probable that the PZT powder was randomly distributed in the polymer and that the shape of the PZT powder must be approximately spherical, as Yamada et al. described in their paper.[8] Conclusions based on the results will be discussed later in this paper.

Theoretical equations derived from the parallel and series models by Newnham et al. are summarized in Table 1. 4.

Table 1. 4 Theoretical equations for piezoelectric d_{33} and d_{31} constants and elastic S_{33} and S_{11} constants of a two phase system (composite) derived from series and parallel models.

Constant	Series Model		Parallel Model	
Piezoelectric d_{33}	$\bar{d}_{33} = \dfrac{{}^1v \cdot {}^1d_{33} \cdot {}^2\varepsilon_{33} + {}^2v \cdot {}^2d_{33} \cdot {}^1\varepsilon_{33}}{{}^1v \cdot {}^2\varepsilon_{33} + {}^2v \cdot {}^1\varepsilon_{33}}$	(T4.1)	$\bar{d}_{33} = \dfrac{{}^1v \cdot {}^1d_{33} \cdot {}^2S_{33} + {}^2v \cdot {}^2d_{33} \cdot {}^1S_{33}}{{}^1v \cdot {}^2S_{33} + {}^2v \cdot {}^1S_{33}}$	(T4.2)
Elastic S_{33}	$\bar{S}_{33} = {}^1v \cdot {}^1S_{33} + {}^2v \cdot {}^2S_{33}$	(T4.3)	$1/\bar{S}_{33} = {}^1v/{}^1S_{33} + {}^2v/{}^2S_{33}$	(T4.4)
Piezoelectric d_{31}	$\bar{d}_{31} = \dfrac{{}^1v \cdot {}^1d_{31} \cdot {}^2\varepsilon_{33} \cdot {}^2S_{11} + {}^2v \cdot {}^2d_{31} \cdot {}^1\varepsilon_{33} \cdot {}^1S_{11}}{({}^1v \cdot {}^2\varepsilon_{33} + {}^2v \cdot {}^1\varepsilon_{33})({}^1v \cdot {}^2S_{11} + {}^2v \cdot {}^1S_{11})}$	(T4.5)	$\bar{d}_{31} = {}^1v \cdot {}^1d_{31} + {}^2v \cdot {}^2d_{31}$	(T4.6)
Elastic S_{11}	$1/\bar{S}_{11} = {}^1v/{}^1S_{11} + {}^2v/{}^2S_{11}$	(T4.7)	$\bar{S}_{11} = {}^1v \cdot {}^1S_{11} + {}^2v \cdot {}^2S_{11}$	(T4.8)

For the parallel model, which Newnham et al. have based on, the dielectric, elastic and piezoelectric constants are expressed as follows:

$$\bar{\varepsilon}_{11} \neq \bar{\varepsilon}_{22}, \; \bar{S}_{11} \neq \bar{S}_{22} \text{ and } \bar{d}_{31} \neq \bar{d}_{32} \tag{1.3}$$

even if

$$\begin{aligned}&{}^1\varepsilon_{11} = {}^1\varepsilon_{22}, \; {}^2\varepsilon_{11} = {}^2\varepsilon_{22}, \; {}^1S_{11} = {}^1S_{22}, \; {}^2S_{11} = {}^2S_{22}, \\ &{}^1d_{31} = {}^1d_{32} \text{ and } {}^2d_{31} = {}^2d_{32}.\end{aligned} \tag{1.4}$$

Here, the author would like to point out that there is another parallel model, where $\bar{\varepsilon}_{11} = \bar{\varepsilon}_{22}$, $\bar{S}_{11} = \bar{S}_{22}$, and $\bar{d}_{31} = \bar{d}_{32}$ when Eq.(1.4) is valid. This parallel model is of the 1-3 type, as shown in Fig. 1.4, which is distinguished from that of the coventional 2-2 type parallel one. The theoretical equations derived from the 1-3 type parallel model are shown in the Table 1.5.

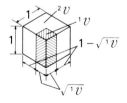

Fig. 1. 4 Schematic representation of the 1-3 type parallel model.

According to the equations shown in the Table 1.5, the d_{31} equation for the 1-3 type PZT (phase 1)-polymer (phase 2) composites should be revised from

$$\bar{d}_{31} = {}^1v \cdot {}^1d_{31} \tag{1.5}$$

to

$$\bar{d}_{31} = \dfrac{{}^1v \cdot {}^1d_{31}}{1 - \sqrt{{}^1v} + {}^1v}, \tag{1.6}$$

assuming that ${}^2S_{11} \gg {}^1S_{11}$ and ${}^2S_{33} \gg {}^1S_{33}$.

Table 1.5 Theoretical equations for the dielectric, elastic and piezoelectric constants of a two phase system (composite) derived from the 1-3 type parallel model.

Constant	Theoretical Equation	
Dielectric ε_{33} and ε_{11}	$\bar{\varepsilon}_{33} = {}^1v \cdot {}^1\varepsilon_{33} + {}^2v \cdot {}^2\varepsilon_{33}$	(T5.1)
	$\bar{\varepsilon}_{11} = \dfrac{\sqrt{{}^1v}}{\dfrac{\sqrt{{}^1v}}{{}^1\varepsilon_{11}} + \dfrac{1-\sqrt{{}^1v}}{{}^2\varepsilon_{11}}} + (1-\sqrt{{}^1v}) \cdot {}^2\varepsilon_{11}$	(T5.2)
Elastic S_{33} and S_{11}	$1/\bar{S}_{33} = {}^1v/{}^1S_{33} + {}^2v/{}^2S_{33}$	(T5.3)
	$\dfrac{1}{\bar{S}_{11}} = \dfrac{\sqrt{{}^1v}}{\sqrt{{}^1v} \cdot {}^1S_{11} + (1-\sqrt{{}^1v}) \cdot {}^2S_{11}} + \dfrac{1-\sqrt{{}^1v}}{{}^2S_{11}}$	(T5.4)
Piezoelectric d_{33} and d_{31}	$\bar{d}_{33} = \dfrac{{}^1v \cdot {}^1d_{33} \cdot {}^2S_{33} + {}^2v \cdot {}^2d_{33} \cdot {}^1S_{33}}{{}^1v \cdot {}^2S_{33} + {}^2v \cdot {}^1S_{33}}$	(T5.5)
	$\bar{d}_{31} = \left\{ \sqrt{{}^1v} \cdot \dfrac{\sqrt{{}^1v} \cdot {}^1d_{31} + (1-\sqrt{{}^1v}) \cdot {}^2d_{31}}{\sqrt{{}^1v} \cdot {}^1S_{11} + (1-\sqrt{{}^1v}) \cdot {}^2S_{11}} + (1-\sqrt{{}^1v}) \cdot \dfrac{{}^2d_{31}}{{}^2S_{11}} \right\} \cdot \bar{S}_{11}$	(T5.6)

Pauer did not give any theoretical expressions for the piezoelectric constants based on the cubes model.

Based on the modified cubes model, Banno et al.[7] developed a theory on the piezoelectric constants of the 0-3 type composite consisting of piezoelectric ceramic particles and synthetic rubber, assuming that

$${}^2S_{11} \gg {}^1S_{11} \text{ and } {}^2S_{33} \gg {}^1S_{33},$$

as expressed in the following equations:

$$\bar{d}_{33} = {}^1d_{33} \cdot \frac{a^3 \cdot [a + (1-a)n]}{a + (1-a)n \cdot ({}^1\varepsilon_{33}/{}^2\varepsilon_{33})} \cdot \frac{1}{\dfrac{(1-a)n}{a + (1-a)n} + a^3} \quad (1.7)$$

$$\bar{d}_{31} = {}^1d_{31} \cdot \frac{a^2 \cdot [a + (1-a)n]^2}{a + (1-a)n({}^1\varepsilon_{33}/{}^2\varepsilon_{33})} \quad (1.8)$$

where $a^3 = {}^1v$.

However this d_{31} expression (Eq.(1.8)) has been derived using the d_{31} equation for the 2-2 type parallel model. As illustrated in Fig. 1.3, the modified cubes model generalizes the 1-3 type (not the 2-2 type) parallel model, cubes model and series models. Accordingly, the d_{31} equation for the 1-3 type parallel model should be used in order to derive the d_{31} equation for the modified cubes model.

This has been revised as follows:[9]

$$\bar{d}_{31} = {}^1d_{31} \cdot \frac{a^2 \cdot [a + (1-a)n]}{a + (1-a)n({}^1\varepsilon_{33}/{}^2\varepsilon_{33})} \cdot \frac{a}{1 - a\sqrt{a + (1-a)n} + a^3}. \quad (1.9)$$

Banno et al. measured the dielectric and piezoelectric d_h constants of the composites consisting of modified $PbTiO_3$ ceramic particles and synthetic rubber. These values were compared to those calculated according to Eq.(T1.6) (Table 1.1) and the following equations:

$$\bar{d}_h = \bar{d}_{33}(\text{Eq.}(1.7)) + 2 \cdot \bar{d}_{31}(\text{Eq.}(1.9)). \qquad (1.10)$$

The results are shown in Fig. 1.5, where the n value for $\bar{\varepsilon}_{33}$ is almost equal to that for $\bar{d}_h$ in a composite. Accordingly, it can be concluded that the theory based on the modified cubes model agreed well with the experimental results.[9]

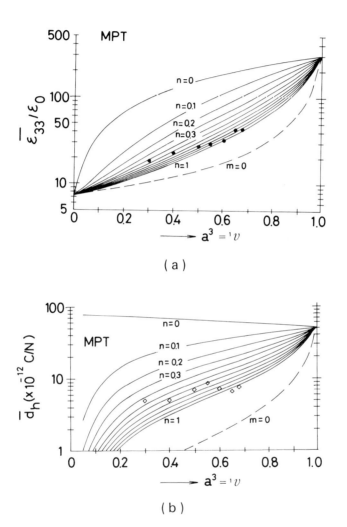

Fig. 1.5 (a) Dielectric constants and (b) piezoelectric d_h constants of composites consisting of modified $PbTiO_3$ ceramic particles and chloroprene rubber as a function of the ceramic volume fraction.

The above-mentioned ellipsoidal ceramic particle arrangement in relation to the plane surface in the PVDF–PZT composites is based on the ellipsoidal model, by Yamada et al.: According to the theory based on the modified cubes model, PZT ceramic particles are not ellipsoidal but spherical, and the unit cell is not cubic but flat tetragonal ($0 < n < 1$, refer to Fig. 1.3).[9]

1.3 Coupling Factor, Frequency Constant and Mechanical Quality Factor of Sandwich-Structure PZT Ceramics

Tashiro et al. have investigated and reported the properties of the sandwich-structure PZT ceramics.[10] They have compared the experimental and calculated values of piezoelectric d_{33} and g_{33} constants according to the equations derived from the series model.

Although they have investigated the coupling factor, frequency constants and mechanical quality factor of sandwich-structure composite ceramics, there has been no comparison between theoretical and experimental values. This is because theoretical equations for the coupling factor, frequency constant and mechanical quality factor based on the series model have not yet been derived.

The frequency constant for the radial mode is expressed by the following equation:

$$f_r \cdot d = \frac{\varphi_1}{\pi} \sqrt{\frac{1}{\rho(1-\sigma^2) \cdot S_{11}^E}} \tag{1.11}$$

If it can be assumed that the density (ρ) and the poisson's ratio (σ) of the sandwich-structure composite ceramics are constant, the frequency constant of the ceramics is expressed as follows:

$$(\bar{f}_r \cdot d)^2 = {}^1v \cdot ({}^1f_r \cdot d)^2 + {}^2v \cdot ({}^2f_r \cdot d)^2. \tag{1.12}$$

The radial coupling factor is expressed by the following equation:

$$k_p^2 = \frac{2 \cdot d_{31}^2}{\varepsilon_{33}^T (S_{11}^E + S_{12}^E)} \tag{1.13}$$

If it can be assumed that the poisson's ratio of the composite ceramics is constant, the radial coupling factor of the composite ceramics k_p can be expressed as a function of the following parameters:

$${}^1v, {}^1\varepsilon_{33}, {}^2\varepsilon_{33}, {}^1f_r \cdot d, {}^2f_r \cdot d, {}^1k_p \text{ and } {}^2k_p.$$

The mechanical quality factor Q_m is defined by the following equations:

$$Q_m = 1/\tan\delta_m = S'/S'' \tag{1.14}$$

where $\tan\delta_m$, S' and S'' are the mechanical loss tangent, real and imaginary parts of elastic constant $\dot{S}(=S' + jS'')$, respectively.

Assuming that

$$({}^1\tan\delta_{m11})^2 \ll 1 \text{ and } ({}^2\tan\delta_{m11})^2 \ll 1, \tag{1.15}$$

theoretical equations for the mechanical quality factor $\overline{Q}_{m33}$ and $\overline{Q}_{m11}$ (or mechanical loss tangent $\overline{\tan\delta}_{m33}$ and $\overline{\tan\delta}_{m11}$), are obtained for the series model as follows:

$$1/\overline{Q}_{m33} = \overline{\tan\delta}_{m33} = \frac{{}^1v \cdot {}^1S_{33} \cdot {}^1\tan\delta_{m33} + {}^2v \cdot {}^2S_{33} \cdot {}^2\tan\delta_{m33}}{{}^1v \cdot {}^1S_{33} + {}^2v \cdot {}^2S_{33}} \tag{1.16}$$

$$1/\overline{Q}_{m11} = \overline{\tan\delta}_{m11} = \frac{{}^1v \cdot {}^2S_{11} \cdot {}^1\tan\delta_{m11} + {}^2v \cdot {}^1S_{11} \cdot {}^2\tan\delta_{m11}}{{}^1v \cdot {}^2S_{11} + {}^2v \cdot {}^1S_{11}} \tag{1.17}$$

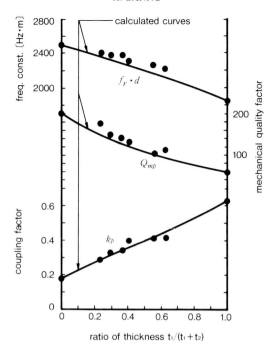

Fig. 1. 6 Comparison between experimental and theoretical values for coupling factor k_p, mechanical quality factor Q_{mp} and frequency constant $f_r \cdot d$ of sandwich-structure PZT ceramics. Theoretical values based on series model.

The theoretical and experimental values measured by Tashiro et al. are compared in Fig. 1. 6. Very good agreements are observed in k_p and Q_{mp}, and a slight disagreement in the frequency constant. This may mean that the density and poisson's ratio of the composite ceramics cannot be assumed to be constant.

1. 4 Porosity Dependencies of Dielectric and Piezoelectric Constants of Polycrystalline Ceramics

Okazaki et al. have investigated the porosity dependencies of the dielectric and piezoelectric constants of the ceramics.[11–15] These are very important data because they are independent of the grain size and only dependent on the porosity of the ceramics.

The modified cubes theory may be applied to experimental data of porous ceramics since they are considered a kind of two phase system composed of bulk ceramic material and pores. In this context, a new parameter, K_S, could be introduced to replace l, m, and n. This parameter would be attributed to the pore shape of phase 1 as shown in Fig. 1. 7. When phase 1 (pores) is cubic or flat tetragonal, K_S becomes unity or less than unity, respectively.

Using K_S and assuming that the relative dielectric constant of the pore is unity, theoretical equations for the dielectric and piezoelectric d_{31} constants of the porous ceramics are obtained as follows:[16]

$$\bar{\varepsilon}_{33} = {}^2\varepsilon_{33} \left\{ 1 + \frac{1}{a({}^2\varepsilon_{33} - 1)K_S^{2/3} + 1} \cdot \frac{a^2}{K_S^{2/3}} - \frac{a^2}{K_S^{2/3}} \right\} \qquad (1.18)$$

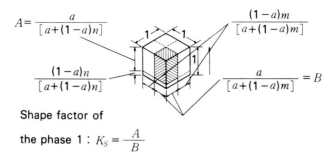

Shape factor of the phase 1 : $K_S = \dfrac{A}{B}$

Fig. 1. 7 Shape factor of phase 1 for the modified cubes model.

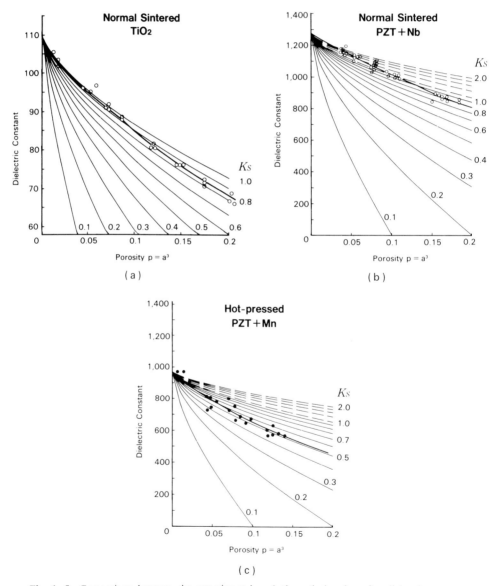

Fig. 1. 8 Comparison between the experimental and theoretical values for dielectric constants of (a) normal sintered TiO_2, (b) normal sintered PZT (Nb added), and (c) hot-pressed PZT (Mn added). Theoretical values obtained according to Eq. (1. 18).[9]

$$\bar{d}_{31} = {}^2d_{31} \left\{ 1 - \frac{a}{K_S^{1/3}} + \frac{\frac{a}{K_S^{1/3}}\left(1 - \frac{a}{K_S^{1/3}}\right)}{1 - a^2 \cdot K_S^{1/3}} \right\} \tag{1.19}$$

where ${}^2\varepsilon_{33}$ and ${}^2d_{31}$ are the relative dielectric and piezoelectric constants of the bulk ceramic material, respectively, and $a^3 (= {}^1v)$ is the volume fraction of the pore. The theoretical and experimental values of the dielectric constants for the normal sintered TiO_2, normal sintered PZT (Nb added) and hot-pressed PZT (Mn added) ceramics, are compared in Figs. 1.8 (a), (b) and (c), respectively. The theoretical and experimental values of piezoelectric d_{31} constants for the hot-pressed PLZT and PZT (Mn-added) ceramics, are compared in Figs. 1.9 (a) and (b), respectively.

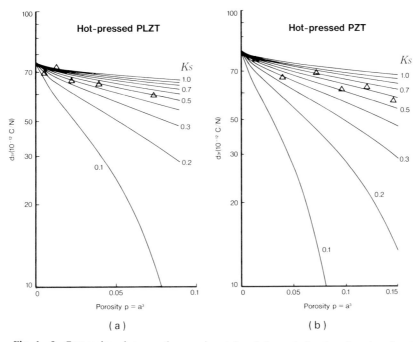

Fig. 1. 9 Comparison between the experimental and theoretical values for piezoelectric d_{31} constants of (a) hot-pressed PLZT and (b) hot-pressed PZT. Theoretical values obtained according to Eq. (1.19).[9]

Using these figures, a remarkable difference can be seen between the K_S values for normal sintered and hot-pressed ceramics. Namely, for the normal sintering, the K_S values are nearly unity, but for hot-pressing they are about 0.5.

From these results, it can be concluded that the pore shape is cubic or spherical for normal sintering, whereas it is flat tetragonal or flat ellipsoidal for hot-pressing.

1.5 Piezoelectric Composite Materials and Their Applications

1.5.1 0–3 Type Composite

The first piezoelectric composite material, made by Kitayama et al. at Nippon

Telegraph and Telephone Public Corporation, was an 0-3 type consisting of polymers and PZT ceramic particles. This composite was made by mixing PVDF (polyvinylidene fluoride), fluorine elastomer, and PZT powder with a mixing roll at 190°C. The composite film was prepared by pressing flakes of the composite, and aluminum was deposited onto both sides of the film for electrodes. Typical characteristics of the composite material, which was polarized in a high DC electric field at 100°C for 2 hours, are shown in Table 1. 6.[17]

Table 1. 6 Typical characteristics of the 0-3 type composite material consisting of polyvinylidene fluoride, fluorine elastomer and PZT ceramic particles. [after T. Kitayama and T. Ueda[17]]

Piezoelectric d_{33} Constant	30×10^{-12} C/N
Dielectric Constant ($\varepsilon_{33}^T/\varepsilon_0$)	118
(tan δ)	0.022
Breakdown Voltage	150 kV/cm
Specific Gravity	5.5 g/cm^3
Elastic Modulus	2.6×10^9 N/m^2
(tan δ)	0.075
Sound Velocity	1.05 (km/sec)

This material has been applied to a "Pizoelectric keyboard".[17] This is made up of a piezoelectric printed circuit board which consists of one printed board and a piezoelectric composite film.[18]

Seo of Mitsubishi Petrochemical Co., Ltd. developed a flexible piezoelectric composite material consisting of polyacetal resin mixed with a polymer having high relative permittivity and piezoelectric ceramic powder.[19] This material has been applied to a impact sensor of the Izod-type impact strength test system.

Inomata et al. of Toshiba Corporation investigated the properties of flexible composite materials cosisting of PVDF, propylene 6-fluorid and $PbTiO_3$-$PbZrO_3$-$Pb(Co_{1/2}W_{1/2})O_3$ ceramic particles. Typical properties of their material are shown in Table 1. 7.[20]

Table 1. 7 Typical characteristics of flexible composite material consising of PVDF, propylene 6-fluoride and $PbTiO_3$-$PbZrO_3$-Pb $(Co_{1/2}W_{1/2})$ O_3 ceramic particles. [after S. Inomata et al.[20]]

Piezoelectric d_{31} Constant	23×10^{-12} C/N
Dielectric Constant ($\varepsilon_{33}^T/\varepsilon_0$)	113
(tan δ)	0.0355
Specific Gravity	5.6 g/cm^3
Elastic Modulus	6.4×10^9 N/m^2

Banno (author of the present paper) et al. investigated the dielectric and piezoelectric d_h constants of the piezoelectric composite material consisting of polychloroprene rubber and $PbTiO_3$ ceramic particles. The results are shown in Fig. 1. 10 as a function of the ceramic volume percent and its microstructure is shown in Figs. 1. 11 (a) and (b).[21]

Banno et al. also developed a piezoelectric co-axial cable using this material as shown in Figs. 1. 12 (a) and (b). One of the features of this cable is that it can be fabricated into a long cable. Applications include hydrophones, probes for measuring the ultrasonic fields in water, blood pressure sensors, guitar and electric piano pick-ups. Two examples of applications are shown in Figs. 1. 13 (a) and (b).

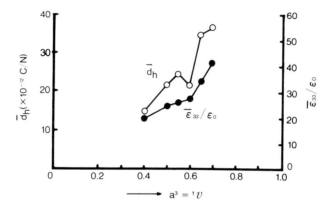

Fig. 1. 10 Piezoelectric d_h constants and dielectric constants of composites consisting of pure lead titanate ceramic particles and chloroprene rubber.

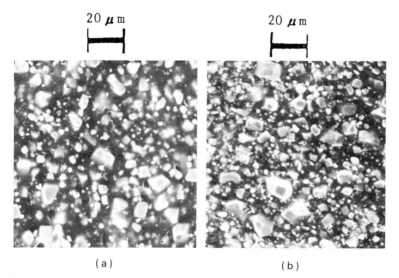

(a) (b)

Fig. 1. 11 SEM micrographs of the composites of pure lead titanate ceramic particles and chloroprene rubber with (a) 40% and (b) 60% ceramic volume.

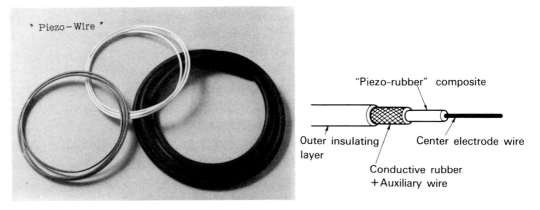

Fig. 1. 12 Photograph and schematic representation of piezoelectric co-axial cable.

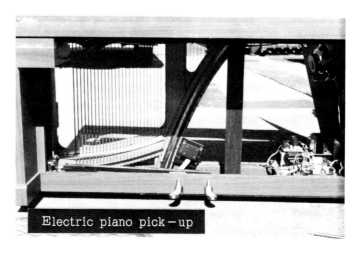

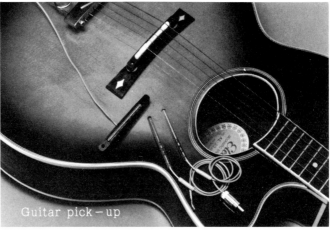

Fig. 1. 13 Electric piano and guitar using piezoelectric co-axial cable as a vibration pick-up.

1.5.2 3-3 Type Composites

In the United States, Skinner et al., of Pennsylvania State University, developed a 3-3 type composite material consisting of PZT and silicone rubber. It possesses a high piezoelectric g_{33}-constant of 300×10^{-3} V·m/N.[22] The techniques which Skinner utilized duplicated the structure of natural coral by the "replamine-form" process. However, the use of the calcium carbonate skeleton of the Goniopora coral as the starting replica material is not particularly compatible with industrial technology.

Miyashita et al. of Toko Corporation proposed a more readily producible type of PZT composite with a regularly controlled ladder structure. Specimens with porosity levels in the 40-60% range also possessed an improved g_{33}-constant of 90×10^{-3} V·m/N.[23] However, the complexity of the ladder structure still poses questions regarding the flexibility of mass producing the composites on a high reliability basis.

Shrout et al. of Pennsylvania State University, also developed a simpler method of fabricating a three-dimensionally interconnected PZT/polymer composite with properties similar to the coral-based composite. The simplified preparation method involves mixing plastic spheres and PZT powder in an organic binder and carefully firing them to form a porous PZT skeleton.[24]

In Japan, Nagata et al. of the National Defence Academy, succeeded in developing a new 3-3 type composite material. It consists of silicone rubber and interconnected porous PZT ceramics prepared by a modified conventional powder sintering. The properties of the interconnected porous PZT ceramics are shown in Table 1.8.[25] A hydrophone using this composite material had nearly four times the output voltage of that using a normal PZT

Table 1.8 Dielectric, elastic and piezoelectric properties of interconnected porous PZT ceramics. [after K. Nagata et al.[25]]

	No. 1	No. 2	No. 3	Dimension
p	0.52	0.17	0.06	porosity
ρ	3.84	6.65	7.52	[g/cm^3]
ε_s	170	1010	1070	
s_{33}^E	108	23	15	[$\times 10^{-12}$ m^2/N]
d_{33}	190	230	220	[$\times 10^{-12}$ m/V]
g_{33}	130	25	23	[$\times 10^{-3}$ V·m/N]
k_{33}	0.49	0.50	0.58	
k_t	0.44	0.45	0.47	
k_p	0.25	0.41	0.48	
Q_m	9	50	140	
$f_r \cdot L$	820	1260	1560	[kHz·mm]
$f_r \cdot d$	1120	1940	2310	[kHz·mm]
T_c	373		371	[°C]
N_i	0.004p	0.001p	0.001p	depolarization factor

Pb(Zr$_{0.53}$-Ti$_{0.47}$)O$_3$ + 0.5 wt.% Nb$_2$O$_5$
No. 1: Interconnected PZT.
No. 2: Normal PZT.

ceramic.

Hikita et al. of Mitsubishi Mining Cement Co., Ltd. investigated the properties of this 3-3 type composite material as a function of ceramic volume percent and as a function of the pore structure of the interconnected porous ceramics.

As the porosity of the ceramic increased, permittivity decreased linearly and the piezoelectric d_{33} constant increased gradually, so that the piezoelectric g_{33} constant increased. The g_{33} value was 200-300 × 10^{-3} V·m/N at a porosity of 0.65.[26]

1. 5. 3 1-3 Type Composite

The dielectric and piezoelectric properties of the 1-3 type composite material have been investigated by Klicker et al. of Pennsylvania State University.[27] This composite material is reported to be applicable to high frequency transducers.[28,29]

Nakaya et al. of Hitachi, Ltd. has investigated the electromechanical properties of the 1-3 type composite fabricated by "dicing and filling" techniques.

Thickness dilatational electromechanical coupling factor k_t of the composite is as high as 0.75 when the PZT volume is 0.25 and the PZT pillar shape (width/thickness) is 0.5-0.6, even if k_t of 100% PZT is about 0.5.

This material has been applied to high frequency (7 MHz) medical ultrasonic probes.[30]

1. 6 Summary

The developments of piezoelectric composites in Japan may be summarized as follows;

(1) The theoretical equations for dielectric and piezoelectric constants were derived from spherical and ellipsoidal models and a modified cubes model which generalizes the 1-3 type parallel, series and cubes models.

(2) The understanding of the ellipsoidal ceramic particle arrangement perpendicular to the plane surface in PVDF-PZT (0-3 type) composites based on the ellipsoidal model, should be revised as follows: the thickness around spherical (not ellipsoidal) ceramic particles in the direction perpendicular to the plane surface is smaller than that in the parallel direction.

(3) Theoretical equations were derived for the series model for the electromechanical coupling factor, frequency constant and mechanical quality factor. They were then applied to sandwich-structure composite ceramics and there was good agreement between theory and experimental results.

(4) The dependencies of dielectric and piezoelectric constants on the shape and percent volume of porous ceramic were theoretically derived from the modified cubes model. The theory was applied to the experimental results obtained by Okazaki et al. and there was good agreement between theory and experimental results.

(5) Developments of piezoelectric composite material were done by Nippon Telegraph and Telephone Public Corporation, Mitsubishi Petrochemical Co., Ltd., Toshiba Corporation and NGK Spark Plug Co., Ltd. for the 0-3 type, Toko Corporation, National Defence Academy and Mitsubishi Mining Cement Co., Ltd. for the 3-3 type

and Hitachi, Ltd. for the 1-3 type.
(6) These materials have been applied to a "Piezoelectric keyboard," impact sensors for an Izod-type strength tester, guitar and piano pick-ups, hydrophones and high-frequency medical ultrasonic probes.

References

1) K. W. Wagner: "Erklärung der Dielektrischen Nachwirkungsvorgänge auf grund Maxwellscher Vorstellungern," Arch. Electrotech., **2** (1914) 371-387 [in German].
2) W.R.Buessem: "Ceramic problems for the consideration of the solid state physicist," The Physics and Chemistry of Ceramics (edited by C. Klingsberg), Gordon and Breach Science Pub. Inc.(1963) 22-29.
3) T.Kitayama and S.Sugawara: "Piezoelectric and pyroelectric properties of polymer-ferroelectric composites," Rep. of Tech. Group, Inst. Electron. Commum. Eng. Japan, **CPM 72**-17 (1972) [in Japanese].
4) L.A.Pauer: "Flexible piezoelectric material," IEEE Int. Conv. Rec.(1973) 1-5.
5) T.Furukawa, K.Fujino and E.Fukada: "Electromechanical properties in the composites of epoxy resin and PZT ceramics," Jpn. J. Appl. Phys., **15** (1976) 2119-2129.
6) R.E.Newnham, D.P.Skinner and L.E.Cross: "Connectivity and piezoelectric-pyroelectric composites," Mater. Res. Bull., **13** (1978) 525-536.
7) H.Banno and S.Saito: "Piezoelectric and dielectric properties of composites of synthetic rubber and $PbTiO_3$ or PZT," Jpn. J. Appl. Phys., **22**, Suppl. 22-2 (1983) 67-69.
8) T.Yamada, T.Ueda and T.Kitayama: "Piezoelectricity of a high-content lead zirconate titanate/polymer composite," J. Appl. Phys., **53** (1982) 4328-4332.
9) H.Banno: "Theoretical equation for dielectric and piezoelectric properties of ferroelectric composites based on modified cubes model," Proc. 6th Int. Meeting on Ferroelectricity (IMF-6, Kobe, Japan, 1985); Jpn. J. Appl. Phys., **24**, Suppl. 24-2 (1985) 243-245.
10) S.Tashiro, N.Arai, H.Igarashi and K.Okazaki: "Piezoelectric properties of the sandwich structured PZT ceramics," Ferroelectrics, **37** (1981) 595-598.
11) K.Okazaki and K.Nagata: "Effects of the density and the grain size of piezoelectric ceramic influencing upon the electrical properties," Trans. IECE Japan, **53**-C (1970)815-822 [in Japanese].
12) K.Okazaki and K.Nagata: "Effects of density and grain size on the elastic and piezoelectric properties of $Pb(Zr-Ti)O_3$ ceramics," Proc. 1971 Int. Conf. on Mechanical Behavior of Materials IV (1972) 404-412.
13) K.Okazaki and H.Igarashi: "Importance of microstructure in electronic ceramics," Proc. 6th Int. Mater. Symp. on Ceramic Microstructure '76 (1976) 564-583.
14) H.Igarashi: "Effects of the porosity of electronic ceramics upon the electrical and magnetic properties," Ceramics Japan, **10** (1975) 799-806 [in Japanese].
15) H.Igarashi: "Control of microstructure of dielectric ceramics and their properties," Ceramics Japan, **12** (1977) 126-131 [in Japanese].
16) H.Banno: "Dependencies of dielectric and piezoelectric constants of polycrystalline ceramics on shape and volume fraction of pore," Preprint of Ann. Meeting Jpn. Ceram. Soc. (1985) 511-512 [in Japanese].
17) T.Kitayama and T.Ueda: "Piezoelectric polymer composite transducer for electronic keyboard," Proc. 1st Meeting on Ferroelectric Materials and their Applications (FMA-1, 1978, FMA office, Kyoto) 263-264.
18) I.Namiki, K.Sugiyama, T.Kitayama and T. Ueda: "Piezoelectric keyboard electric design condition," IEEE Trans. Components, Hybrids & Manuf. Technol., **CHMT**-4 (1981) 304-310.
19) I.Seo: Piezo-electrical Material, U.S. Patent 4,128,489 (1978).
20) S.Inomata, Y.Fujimori, N.Kaneko, Y.Yamashita and Y.Seo: "Composite type piezoelectric materials and their application,"

Rep. of Tech. Group, Inst. Electron. Commun. Eng. Japan, **CPM** 82-30 (1982) 63-70 [in Japanese].

21) H. Banno: "Recent developments of piezoelectric ceramic products and composites of synthetic rubber and piezoelectric ceramic particles," Ferroelectrics, **50** (1983) 3-12.

22) D.P.Skinner, R.E.Newnham and L.E.Cross: "Flexible composite transducers," Mater. Res. Bull., **13** (1978) 599-607.

23) M.Miyashita, K.Takano and T.Toda: "Preparation and properties of PZT ceramics with ladder type structure," Ferroelectrics, **28** (1980) 397-400.

24) T.R.Shrout, W.A.Schulze and J.V.Biggers: "Simplified fabrication of PZT/polymer composites," Mater. Res. Bull., **14** (1970) 1553-1559.

25) K.Nagata, H.Igarashi, K.Okazaki and R.C. Bradt: "Properties of an interconnected porous Pb(Zr,Ti)O$_3$ ceramic," Jpn. J. Appl. Phys., **19** (1980) L37-40.

26) K.Hikita, K.Yamada, M.Nishioka and M. Ohno: "Effect of porous structure to piezoelectric properties of PZT ceramics," Jpn. J. Appl. Phys., **22**, Suppl. 22-2 (1983) 64-66.

27) K.A.Klicker, J.V.Biggers and R.E.Newnham: "Composites of PZT and epoxy for hydrostatic transducer applications," J. Am. Ceram. Soc., **64** (1981) 5-9.

28) L.J.Bowen and T.R.Gururaja: "High-frequency electromechanical properties of piezoelectric ceramic/polymer composites in broadband applications," J. Appl. Phys., **51** (1980) 5661-5666.

29) T.R.Gururaja et al.: "Composite piezoelectric transducers," IEEE 1980 Ultrasonics Symp. Proc. (1980) 576-581.

30) C.Nakaya, H.Takeuchi and K.Katakura: "Medical ultrasonic transducer using PZT/polymer composites," Rep. of Tech. Group, Inst. Electron. Commun. Eng. Japan, **US83**-30 (1983)[in Japanese].

2. Research Trends in Ceramic Sensors

Noboru ICHINOSE*

Abstract

Various ceramic sensors have been recently developed. Because ceramic materials excel over single crystals and thin films in physical and chemical stability as well as mass producibility, they can be utilized for a wide range of applications, such as ceramic sensors. Typical caramic sensors are as follows: (1)Temperature sensors such as NTC, PTC and critical temperature resistors. (2)Pressure sensors which utilize the piezoelectric characteristics of ceramics. (3)Gas and humidity sensors which detect adsorption and desorption of gases by changes in electrical conductivity. (4)Infrared ray sensors which utilize the pyroelectric characteristics of ceramics.

Keywords: ceramic materials, temperature sensor, gas sensor, humidity sensor, infrared ray sensor.

2.1 Introduction

Recently, there have been remarkable developments in electronics technology. There has been progress in ceramic materials paralleling that in electronics. Many new materials and manufacturing techniques have been developed to satisfy the demanding requirements of general users. Ceramic materials development has been accelerated, and is now being fed back into the electronics industry.

This paper discusses how ceramic materials are utilized as sensors.[1] Further descussion of sensors might require a more exact definition, but for this paper, sensors should simply be considered "detecting elements." Ceramic semiconductor materials are useful as components for sensors.

Ceramic semiconductor materials, currently in use, can be classified as follows:
(1) Those that utilize the physical properties of the grain itself.
(2) Those that utilize the properties of the grain boundary.
(3) Those that utilize surface effects.

Functional devices developed by utilizing special ceramics characteristics are shown in Table 2.1. Temperture sensors applying semiconductive, dielectric or magnetic properties; infrared sensors using the pyroelectric effect; and ultrasonic sensors using the piezoelectric effect, are all found in category (1). Despite the fact that these devices all have

---

* Department of Metallurgical Engineering, School of Science and Engineering, Waseda University, 3-4-1, Okubo, Shinjuku-ku, Tokyo 160, Japan.

Table 2.1 Classification of ceramic semiconductor.

(1) Physical Properties of Grain Itself
   (a) NTC thermistor
   (b) High temperature thermistor
   (c) Oxygen gas sensor

(2) Properties of Grain Boundary
   (a) PTC thermistor
   (b) Semiconducting capacitor
   (c) ZnO varistor

(3) Surface Effects
   (a) $BaTiO_3$ varistor
   (b) Gas sensor. Humidity sensor
   (c) Catalyst

the same fine properties, some utilize only the surface layer near the grain boundary rather than the overall bulk.

Examples of these devices are Positive Temperature Coefficient (PTC) thermistors, in which the resistance value changes sharply with sharp changes in temperature, and non-linear resistors (varistors), in which current flow changes rapidly with fluctuations in voltage. In all devices a junction layer is formed within the semiconductor bulk surface layer utilizing the abnomally fast grain boundary diffusion achieved when sintering to diffuse or deposit impurities.

With porous ceramics, vapors and gases pass through pores, diffuse in the ceramics, and are adsorbed on the crystal surface. Since there is a close relation, particularly for semiconductor ceramics, with the surface layer, electrical conductivity usually changes sharply with moisture or gas adsorption. Therefore, since ceramics possess such structural features as bulk properties, grain boundary, surface and pores, they are suitable as sensor material for use in gaseous and humid atmospheres.

Ceramic materials such as those shown in Table 2.2 are applicable for sensors. A detailed explanation is provided here for only a few of them.

2.2 Gas Sensors

Generally speaking, sensors that are capable of directly giving output as electrical signals are suitable for the detection of gas leakage. Such sensors are roughly classified into semiconductor and contact combustion types.

The semiconductor type operates in the following way. If a semiconductor (generally, n-type oxide semiconductors such as SnO_2, Fe_2O_3 and ZnO), which is heated to a high temperature, comes into contact with combustible gas, its electric resistance is lowered. This semiconductor property is utilized for the semiconductor type gas sensor.

The contact combustion type detects gas in the following way: If combustible gas reacts with a catalyst and burns on a heated platinum wire, the temperature rise of the wire is detected as an increase in its electrical resistance.

Table 2.2 Examples of ceramics used as sensors.

Output		Effect		Material (form)	Remarks
Temperature sensor	Change in resistance	Temperature changes carrier density	(NTC)	NiO, FeO, CoO, MuO, CoO-Al$_2$O$_3$, SiC, (Bulk, thick film, thin film)	Thermometer, bolometer
			(PTC)	Semiconductive BaTiO$_3$ (sintered body)	Overheat protection sensor
		Semiconductor–metal phase transition		VO$_2$, V$_2$O$_5$	Thermal switch
	Change in magnetism	Ferrimagnetism–paramagnetism transition		Mn-Zn based ferrite	Thermal switch
	Electromotive force	Concentrated oxygen cell		Stabilized zirconia	High-temperature corrosion-resistance thermometer
Position/speed sensor	Change in waveform of reflected wave	Piezoelectric effect		PZT : (PbTiO$_3$-PbZrO$_3$)	Fish finder, flaw detector, blood flow meter
Opto sensor	Electromotive force	Collecting effect		LiNbO$_3$, LiTaO$_3$, PZT, SrTiO$_3$	Infrared ray detection
	Visible light	Anti-stoke measurement		LaF$_3$ (Yb, Er)	Infrared ray detection
		Progressive wave doubling effect		Piezoelectrics Ba$_2$NaNb$_5$O$_{15}$ (BNN), LiNbO$_3$	
		Fluorescence		ZnS (Cu, Al), Y$_2$O$_2$S (Eu)	Color TV CRT
				ZnS (Cu, Al)	X-ray monitor
		Thermal fluorescence		CaF$_2$	Thermal fluorescence dosimeter
Gas sensor	Change in resistance	Inflammable gas contact combustion reaction heat		Pt catalyst/ alumina /Pt line	Inflammable gas thermometer and alarm
		Migration of charge due to adsorption and desorption of gas from an oxidized semiconductor		SnO$_2$, In$_2$O$_3$, ZnO, WO$_3$, γ-Fe$_2$O$_3$, NiO, CoO, Cr$_2$O$_3$, TiO$_2$, LaNiO$_3$, (La, Sr) CoO$_3$, (Ba, La)TiO$_3$, etc.	Gas alarm
		Temperature changes in a thermistor due to gas heat conductive dissipation		Thermistor	High-density gas sensor
		Stoichiometric change in oxide semiconductors		TiO$_2$, CoO-MgO	Automobile exhaust gas sensor
	Electromotive force	High-temperature solid electrolytic concentrated oxygen cell		Stabilized zirconia (ZrO$_2$-CaO, ZrO$_2$-Y$_2$O$_3$, La$_2$O$_3$) Thoria (ThO$_2$ Y$_2$O$_3$)	Exhaust gas sensor (random sensor) Molten steel and molten steel oxygen content analyzer
	Amount of electricity	Coulomb titration		Stabilized zirconia	Lean combustion oxygen sensor
Humidity sensor	Change in resistance	Moisture absorption in conductivity		LiCl, P$_2$O$_5$, ZnO-Li$_2$O	Hygrometer
		Oxide semiconductor		TiO$_2$, NiFe$_2$O$_4$, MgCr$_2$O$_4$, MgCr$_2$O$_4$+TiO$_2$, ZnO Ni ferrite, Fe$_3$O$_4$, Colloid	Hygrometer
	Permittivity	Change in permittivity due to moisture absorption		Al$_2$O$_3$	Hygrometer
Ion sensor	Electromotive force	Solid electrolytic film concentrated cell		AgX, LaF$_2$, Ag$_2$S$_3$, Thin glass film, CdS, AgI	Ion concentration sensor
	Resistance	Gate adsorption effect MOSFET		Si (Gate material H$^+$ for:Si$_3$N$_4$/SiO$_2$, S^{2-} for: Ag$_2$S, X$^-$ for: AgX, PbO)	Ion sensitivity FET (IFSET)

Speaking of the merits and demerits of these two types, it may be noted that the semiconductor type is highly sensitive to low concentrations of gas and that the output of the contact combustion type is proportional to the gas concentration. Both types are required to meet the following conditions:
(1) They should be highly sensitive to the gases to be detected.
(2) They should not detect any gases other than those to be detected.
(3) They are required to maintain stable operation characteristics over a long period of time.

Importance is to be attached to the above-mentioned conditions in developing a gas sensor. It may be pointed out that many of the recently developed gas sensors are considerably improved in these characteristics. Some of these are outlined below.

A highly selective gas sensor is obtained by adding a catalyst to the ZnO oxide semiconductor.[2] Its construction is given in Fig. 2. 1. Of the ZnO type sensors, those using a catalyst containing a Pt compound, are highly sensitive to hydrocarbon gases, and those using a Pd compound, to CO and H_2. Figure 2. 2 shows the characteristics of ZnO (Pt) in detecting gases of different concentrations. As this figure shows, it is highly sensitive to isobutane (i-C_4H_{10}) and propane (C_3H_8) and is used as a sensor to detect propane gas.

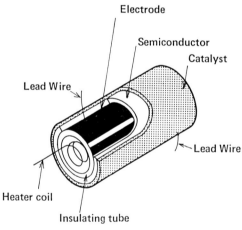

Fig. 2. 1   Structure of element.

As shown in Fig. 2. 1, a catalyst layer surrounds the semiconductor layer. Experiments thus far have shown the role of the catalyst layer to be as follows:
(a) The sensitivity of the ZnO semiconductor to gases (such as isobutane) does not differ much according to its composition, and the sensitivity is rather low. Its stability, however, particularly its aging characteristics and its initial current passage stability varies widely depending on its composition.
(b) The resistance value of the ZnO device with a catalyst layer rise sharply in air, but hardly changes in gas. A different gas selectivity can be obtained from the same semiconductor by changing its catalyst layer.
(c) The mutual relation between the semiconductor layer and the catalyst layer does not depend only on the contact between the two layers. The two layers interact even if

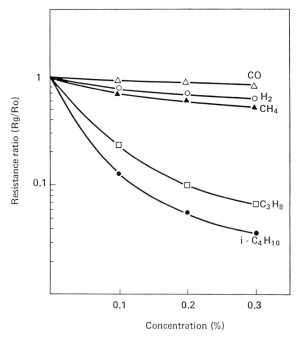

Fig. 2. 2 Sensitivity of modified ZnO gas sensor.

they are separated from each other to a certain extent.

(d) By providing a catalyst layer, excessive Zn ions of the ZnO are adsorbed into hydrogen. In other words, the catalyst layer promotes the adsorption of hydrogen from the air into the semiconductor.

Liquid propane gas (LPG) is commonly used today and accidents caused by LPG are increasing. Accordingly, the demand for gas leak alarms is increasing. In Japan, 18 million households use LPG, and to prevent LPG explosions, the development of high-performance ceramic sensors is being accelerated. A recently developed ZnO element has excellent sensitivity, separation and stability, and matches present requirements. Its gas selectivity is shown in Fig. 2. 2.

2. 3 Humidity Sensors

As the detection of humidity becomes increasingly systems-oriented, there is a strong demand for humidity detection by means of a sensing device that utilizes electrical signals, particularly fluctuations in impedance. There have been many proposals for humidity sensors but no definitive ones have yet been devised.

The ceramic humidity sensors recently developed, are those such as $MgCr_2O_4$-TiO_2, ZnO-Cr_2O_3 and TiO_2-V_2O_5. Among the conditions a humidity sensor is required to meet are: quick response, precise reproducibility, changes in resistance values within a practical range of actual measurement, a broad humidity response range, resistance to dirt and other gases, stability and excellent aging characteristics.

A chemically stable ceramic material consisting mainly of ZnO has recently been developed.[3] With an activated surface, it can measure humidity stably and continuously.

Unlike the $MgCr_2O_4$-TiO_2 sensor, it does not require cleaning by means of a heater. Furthermore, it lends itself more easily to mass production so that it can be offered at a lower price. The small size sensor measures 8.5 mm in diameter and 0.25 mm in thickness.

The construction of the sensor is as shown in Fig. 2. 3. Porous electrodes are baked on to a porous ceramic disc, platinum-rhodium wires are welded to the electrodes, and the entire device is hermetically sealed. A lead wire is then welded to it. It is then mounted in a square plastic case with mesh-filter ventilation holes, and fixed with resin to withstand vibration and fall tests.

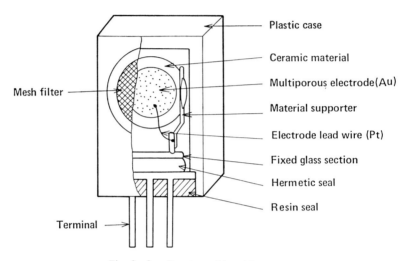

Fig. 2. 3 Structure of humidity sensor.

The humidity-sensing principle and mechanism are as follows: The cross-sectional construction of the ceramic substrate is as shown in Fig. 2. 4. The ceramic substrate represents a porous spinel construction consisting of $ZnCr_2O_4$ with a grain size of 2-3 μm, and the surfaces of these particles are evenly covered with a thin glazed film of $LiZnVO_4$ metal oxide. This thin film layer contains the humidity sensing Li-O, and is a stable

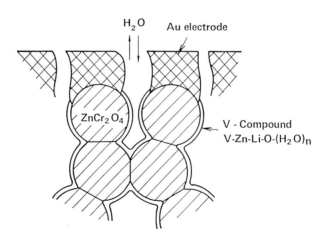

Fig. 2. 4 Sensor cross-section diagram.

humidity-sensing layer as it is fixed in the V-O matrix construction. The surface of this humidity-sensing layer has a stable OH radical, on which an adsorbed layer—a multimolecular layer of water molecules—is formed to promote conductivity.

The $ZnO-Cr_2O_3$ sensor is designed for continuous humidity detection over a long period of time. It was tested for its resistance to ambient conditions, such as aggressive gases, smoke, dust, oils and aging due to air-pollution, and it was found to exhibit excellent characteristics in these respects. Its main characteristics are shown in Figs. 2. 5 and 2. 6.

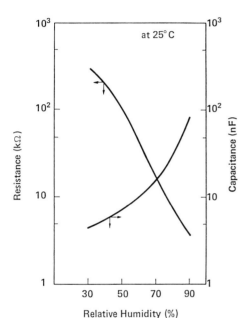

Fig. 2. 5 Typical relative humidity-resistance and relative humidity-capacitance characteristics.

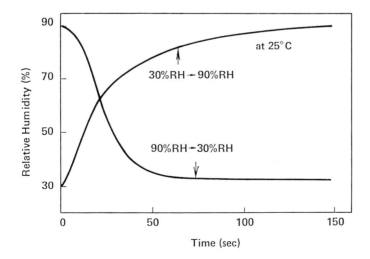

Fig. 2 6 Humidity response characteristics.

2.4 Pyroelectric Infrared Sensors

Pyroelectric infrared sensors have been widely used in consumer electric appliances and other electric equipment, such as microwave ovens, automatic door systems, human body sensors and burglar alarm systems. Pyroelectric infrared sensors have good sensing characteristics. Unlike photon sensors, they can be used at room temperature and their infrared response does not depend on the wavelength of the infrared rays used. The performance of the pyroelectric materials is represented by the figure of merit F_v, defined as $P/\varepsilon C_v$. Here, P, ε and C_v are the pyroelectric coefficient, relative dielectric constant and volume specific heat, respectively. Nowadays, there are various kinds of pyroelectric materials commercially available; for example, $LiTaO_3$ single crystals. Another of these are $PbTiO_3$ ceramics which have been regarded as good pyroelectric materials because of their large pyroelectric coefficient P, small dielectric constant ε and high Curie temperature T_c.

In this paper, the pyroelectric properties of $(Pb, Ca)[(Co_{1/2}W_{1/2})Ti]O_3$ ceramic have been discussed.[4] This pyroelectric ceramic has been used in infrared sensors. These infrared sensors are assembled by hybrid IC technology. A new low temperature fired ceramic $(BaSn(BO_3)_2)$ has been used as the substrate, because of its low thermal conductivity.[5]

The temperature dependence of the pyroelectric coefficient P of $(Pb_{1-x}Ca_x)[(Co_{1/2}W_{1/2})Ti]O_3$ ceramic is shown in Fig. 2.7. It is important for practical application to minimize any variation in pyroelectric coefficient P that might be caused by an increase in ambient temperature. In the case of from $x=0.05$ to $x=0.28$, pyroelectric coefficient P is hardly dependent on the temperature increase from room temperature to 100°C.

Pyroelectric coefficient P, relative dielectric constant ε, and figure of merit F_v, for $(Pb_{1-x}Ca_x)[(Co_{1/2}W_{1/2})Ti]O_3$ are shown in Fig. 2.8 as a function of Ca concentration. The

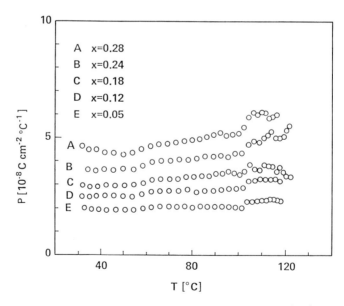

Fig. 2.7 The temperature (T) dependence of the pyroelectric coefficient (P) of $(Pb_{1-x}Ca_x)[(Co_{1/2}W_{1/2})_{0.04}Ti_{0.96}]O_3$ for increasing temperatures.

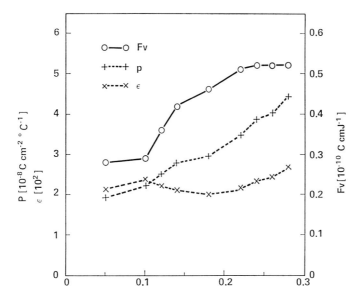

Fig. 2. 8 The composition (x) dependence of the pyroelectric coefficient (P), the dielectric coefficient (ε) and the pyroelectric figure of merit (F_v) of $(Pb_{1-x}\ Ca_x)[(Co_{1/2}W_{1/2})_{0.04}Ti_{0.96}]O_3$.

introduction of Ca into PbTiO$_3$ ceramic increases the pyroelectric coefficient to a great extent. The pyroelectric coefficient reaches 4.43×10^{-8} Ccm^{-2}°C^{-1} with a Ca concentration of $x = 0.28$. Pyroelectric coefficient P for PbTiO$_3$ ceramic containing La and Mn is 1.8×10^{-8} Ccm^{-2}°C^{-1}. Pyroelectric coefficient P for (Pb, Ca)[(Co$_{1/2}$W$_{1/2}$)Ti]O$_3$ has been improved to twice that for the representative PbTiO$_3$ ceramic.

The change in the relative dielectric constant is markedly small. As Ca is introduced into the host lattice, the relative dielectric constant decreases gradually up to a Ca concentration of $x = 0.18$. From $x = 0.18$ to $x = 0.30$, the dielectric constant increases gradually. Above $x = 0.30$, it increases rapidly. Another representative pyroelectric ceramic, PZT, has a large dielectric constant 380–1,800; (Pb, Ca)[(Co$_{1/2}$W$_{1/2}$)Ti]O$_3$ ceramic has a low relative dielectric constant, in the neighborhood of 200. Figure of merit F_v increases in the Ca concentration range from $x = 0.05$ to $x = 0.24$. From $x = 0.24$ to $x = 0.28$, F_v reaches a high value of 0.61×10^{-10} CcmJ^{-1}. At present, pure PbTiO$_3$ ceramic cannot be obtained, because of difficulty in sintering. Therefore, compared with representative PbTiO$_3$ ceramic containing La and Mn, F_v has been greatly improved from 0.3×10^{-10} CcmJ^{-1} to 0.61×10^{-10} CcmJ^{-1}.

As mentioned above, pyroelectric characteristics have been improved to twice the values of modified PbTiO$_3$ ceramics. To demonstrate the improvement of these pyroelectric properties, spontaneous polarization P_S was measured by the Sawyer-Tower method.[6] Ca concentration dependence of P_S for the $(Pb_{1-x}Ca_x)[(Co_{1/2}W_{1/2})_{0.04}Ti_{0.96}]O_3$ system is shown in Fig. 2. 9. Spontaneous polarization P_S increases in the Ca concentration range from $x = 0.05$ to $x = 0.30$ and reaches a high of 42 μC/cm^2 at $x = 0.30$. This value is half that of single crystal PbTiO$_3$. This result shows that improvement of pyroelectric properties is due to P_S

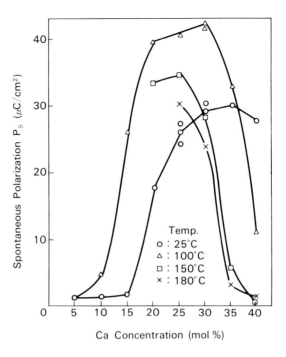

Fig. 2. 9 Ca concentration dependence of spontaneous polarization P_s for $(Pb_{1-x}Ca_x)[(Co_{1/2}W_{1/2})_{0.04}Ti_{0.96}]O_3$ system.

variation increase in the room temperature to 100°C range. The temperature variation results from Curie temperature decrease caused by the increase in Ca concentration.

It has been found that the introduction of Ca into $PbTiO_3$ ceramic exerts a great influence on the $PbTiO_3$ ceramic pyroelectric properties.

The chopping frequency dependence of responsivity R_v is shown in Fig. 2. 10 as a function of load resistance. The maximum responsivity of the peak shifts toward the frequency range below 1 Hz, and further increases to a great extent. The resistance of the pyroelectric element has an appropriate value of 1×10^{11} Ω. Therefore, the sensor can operate without the load resistance. This helps reduce production costs, and simplities the wiring. Human body sensing requires a high responsivity value in the frequency range below 1 Hz. This sensor is remarkably applicable for use as a human body sensor.

The sensor characteristics are summerized in Table 2. 3. Principally, pyroelectric infrared response is not dependent upon the wavelength of the infrared rays used. However, the response is affected by window materials. The sensor with a cut-off filter less than 7 μm thick has a high responsivity value of 780 V/W and a relative detectivity D^* of 1.1×10^8 cm$\sqrt{Hz}$/W at 1 Hz. Because of the transmission difference between window materials, the responsivity and relative detectivity for a sensor with a Si window was reduced to 550 V/W, and 0.8×10^8 cm$\sqrt{Hz}$/W, respectively.

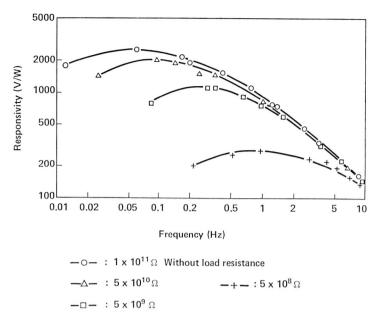

−◦− : 1 × 10¹¹ Ω Without load resistance
−△− : 5 × 10¹⁰ Ω   −+− : 5 × 10⁸ Ω
−□− : 5 × 10⁹ Ω

Fig. 2. 10 Chopping frequency dependence of responsivity as a function of lead resistance.

Table 2. 3 Performance factors of pyroelectric sensors using various pyroelectric materials.

	Optical Wavelength (μm)	R_v (V/W)	NEP (W/$\sqrt{Hz}$)	D^* (cm$\sqrt{Hz}$/W)	A (mm)
TGS (Barnes)	2–20	2400 (15Hz)	2.5×10^{-10} (500,15,1)	4.0×10^8 (500,15,1)	1×1
LiTaO$_3$ (Sanyo)	1–15	400 (10Hz)	6.0×10^{-10} (500,10,1)	3.3×10^8 (500,10,1)	2×2
PbTiO$_3$ (Matsushita)	2–15	45 (20Hz)	1.0×10^{-9} (500,20,1)	1.0×10^8 (500,20,1)	1×1
PZT (Murata)	2–20	150 (6Hz)	1.8×10^{-9} (700,6,0.5)	1.0×10^8 (700,6,0.5)	2ϕ

2. 5 Applications of Ceramic Sensors

Outlined above are representative ceramic sensors used as gas sensors, humidity sensors and pyroelectric sensors. Their applications are not yet very extensive.

Semiconductor gas sensors are characterized by a quicker response to gases than gas sensors of other types, and are highly sensitive to gases of low concentrations (3,000 ppm or less). But they have the following disadvantages: They are lacking in reliability, and it is difficult to manufacture devices meeting certain specifications. They are also lacking in gas-selectivity and are liable to be affected by various kinds of gases.

Semiconductor gas sensors were first put to practical use in Japan. They were almost always used for gas-leakage alarms. It is already 10 years since the first gas sensor of this type was put on the market. In the meantime, it was discovered that these sensors required much improvement. There has not yet been enough basic study of semiconductor gas sensors.

At present, we have several models explaining changes in conductivity due to the adsorption of gases, but there has been no significant study of the interaction between the surface of the semiconductor and gas molecules, including the mechanism of activation and the adsorption phenomenon. Active studies are being made on ways of obtaining a defect-free surface, but it is not yet possible to sufficiently control the surface. In order to improve the reliability of gas sensors, it is essential to control the surface and remove defects. When this problem is overcome, new sensors for LP gas and city gas will be developed.

As shown in Table 2.4, there are many potential application areas for humidity sensors. But they are not so widespread in actual use. One of the reasons for this is that they are not used extensively for home electrical appliances. Among the dry-type room air conditioners introduced in 1982, there were some models utilizing ceramic humidity sensors. They were improved in humidity-sensing accuracy and were capable of fine

Table 2.4 Application areas of humidity sensors by industry: operating temperature and humidity ranges.

Industry	Application Areas	Operating Temperature and Humidity Ranges		Remarks
		Temperature (°C)	Humidity (% RH)	
Home Electric Appliances	Air conditioners	5-40	40-70	Air-conditioning
	Dryers	80	0-40	Drying of clothing
	Microwave ovens	5-100	2-100	Cooking control
	VTRs	-5-60	60-100	Dew-prevention
Automobile	Rear window defogger	-20-80	50-100	Dew prevention
Medical	Medical treatment equipment	10-30	80-100	Respiratory equipment
	Incubators	10-30	50-80	Air-conditioning
Industry	Textile	10-30	50-100	Spinning
	Dryers	50-100	0-50	Ceramics, timber drying
	Powder humidity	5-100	0-50	Ceramic material
	Dried foodstuff	50-100	0-50	
	Electronic parts	5-40	0-50	Magnetic heads, LSIs, ICs
Agriculture, Forestry, Stockbreeding	Greenhouse air-conditioning	5-40	0-100	Air-conditioning
	Dew prevention in tea-leaf growing	-10-60	50-100	Dew prevention
	Broiler farming	20-25	40-70	Health control
Measurement	Thermostatic bath	-5-100	0-100	Precision measurement
	Radiosonde	-50-40	0-100	Precision meteorological measurement
	Hygrometer	-5-100	0-100	Control recorders
Others	Soil humidity			Plant growing, landslides

adjustments of humidity. The consumption of humidity sensors will increase rapidly, if they are used more for consumer products.

In developing sensors, it is important to accommodate user needs. Many of the sensors recently developed are not in line with user requirements. As a result, users are only left with a feeling of uncertainly about sensors, and users tend to be conservative.

If users no longer feel uncertain about the use of sensors, and if higher quality sensors are developed that meet user requirements, their applications will expand remarkably.

2.6 Conclusion

Since there are many types of ceramics material, their functions are also diverse. Material performances demanded by sensor performance are shown in Table 2.5. Although ceramic material cannot meet the requirements of all sensors, many sensors, such as those shown in Table 2.2, can be realized utilizing ceramic multiplicity. The superior ceramic functions are now beginning to find use in sensors.

The properties of ceramics, which are polycrystals, have not been sufficiently studied. There is a gap between science and technology as far as ceramics are concerned. Therefore,

Table 2.5 Material performance required by sensor performance.

	Sensor performance	Perfomance demanded of sensor materials	Remarks
Functional characteristics	Detecting range/dynamic range Sensitivity/detecting limit	Must be wide	According to the fundamental characteristics of the material
	Response speed/frequency	Response speed to be high in the principal process	
	Selectivity	Must have superior selective functions	
	Precision	Correction and compensation of output characteristics should be easy	Linear output, logarithm output
Reliability	Accuracy	Long term stability of material	(Drift and hysteresis to be below the limit)
	Overload resistance	Must not break down or change in quality relative to excessive input	Impact load
	Durability	Must not breadk down mehanically or thermally while handling	Impact, vibration, thermal shock, heat cycle
	Environment resistance	Must not change in quality with usage environment	High temperature, low temperature, high moisture
	Life	Must be minimum fatigue, wear, and change in quality	Atmosphere gas, contamination, vibration
Economic characteristics	Maintainability — Interchangeability	Must have good manufacturing reproducibility of material characteristics	
	Maintainability — Preservability	Must be no change in quality when storing for extended periods	
	Maintainability — Maintenance costs	Contamination, interference and change in quality must not occur easily	
	Manufacturing characteristics — Manufacturing yield	Manufacturing reproducibility of material characteristics must be good	
	Manufacturing characteristics — Manufacturing process cost	Manufacturing processes must be simple and easy to control	
	Manufacturing characteristics — Material cost	Must avoid high-priced raw materials	

because "characterization" will fill this gap, we cannot overlook "characterization" in the development of ceramic materials. It will be necessary to thoroughly study and explore this method to systematically develop materials.

The complexity of polycrystals due to micro-structures has tended to cause studies in ceramics to become empirical. However, there has been a recent trend toward a more precise control of micro-structures for utilizing the characteristic features of sintered materials.

The importance of utilizing the grain boundary is recognized in producing boundary layer (BL) capacitors and ZnO varistors. To attain composite micro-structures with excellent producibility is a major objective in the future development of ceramic materials. Therefore, new electronic ceramics will be developed during these studies.

Previously, most ceramics were oxides. However, for sensor devices of the future it will be necessary to attach more importance to Si_3N_4, AlN and other nitrides, and SiC and other carbides.

References

1) N.Ichinose : "Trends in ceramic sensors," J. Electron. Eng., **19** (Oct. 1982) 81-84.
2) M.Katsura, M.Shiratori, T.Takahashi, Y. Yokomizo and N.Ichinose : "Catalyst effect on zinc oxide semiconductor gas sensor," Proc. Int. Meeting on Chemical Sensors, Kodansha-Elsevier (1983) 101-106.
3) S.Uno, M.Harata, H.Hiraki, K.Sakuma and Y.Yokomizo : "$ZnCr_2O_4$-$LiZnVO_4$ ceramic humidity sensor," Proc. Int. Meeting on Chemical Sensors, Kodansha-Elsevier (1983) 357-380.
4) M.Nakamoto, Y.Hirao, Y.Yamashita and N. Iwase : "Pyroelectric infrared sensor using (Pb,Ca)$[(Co_{1/2}W_{1/2})$ Ti$]O_3$ ceramic," Proc. 4th Sensor Symposium, Tokyo (1984) 209-212.
5) T. Dinh Thank, N. Iwase, H. Egami and E. Echimori : "Low temperature sintered ceramics for hybrid functional circuit (HFC) substrates," IMC 1984 Proceeding, Tokyo 220 (1984).
6) N. Ichinose, Y.Hirao, M.Nakamoto and Y. Yamashita : "Pyroelectric infrared sensor using modified $PbTiO_3$ and its applications," Jpn. J. Appl. Phys., **24**, Suppl. 24-3 (1985) 178-180.

3. Organic-Intercalation Compounds

Fumikazu KANAMARU*

Abstract

Interactions of organic substances with solid surfaces of inorganic layered substances are spotlighted in the material sciences field. Three kinds of organic-intercalation compounds are illustrated using several selected examples: stereospecific synthesis of coordination polymers in the interlayer of host substances, effects of organic-intercalation on the properties of host substances and grafting of organic molecules onto the interlamellar surface of host substances.

Keywords: organic intercalation compound, charge transfer type intercalation compound, stereospecific polymerization in the interlayer of inorganic layered substance, organic derivative of inorganic layered substance.

3.1 Introduction

Many layered inorganic compounds, in which interlayer forces are much weaker than those in the layer, react with various kinds of atoms and molecules (guest species or intercalants), by absorbing them in the interlayer space without destroying the layer's atomic arrangement. A reaction in which the guest species are inserted into the interlayer region of the layered compounds (host compounds) is called "intercalation" and the resultant products are called "intercalation compounds."

Graphite intercalation compounds which are among the most typical intercalation compounds have been extensively studied. Graphite can intercalate a large number of intercalants (atoms, ions and molecules). Some intercalants such as alkaline atoms act as electron donors while other intercalants act as electron acceptors, as seen in the case of $FeCl_3$-bearing graphite.

On the other hand, intercalation compounds which incorporate organic molecules as intercalants have been investigated in clay-organic systems. Such studies have been done largely in the areas of soil science, earth science, catalystic reactions and structural chemistry. Structural analysis of clay-organic compounds by X-ray diffraction revealed that the organic molecules absorbed into the interlayer between silicate layers assume a rather regular arrangement depending on crystallographic symmetry of the silicate layer.[1-3] The host-guest interaction has been also studied by IR, Mössbauer effect and ESR measurements[4-10] in relation to stability of the intercalation compounds and to arrange-

---

* The Institute of Scientific and Industrial Research, Osaka University, 8-1, Mihogaoka, Ibaragi, Osaka 567, Japan.

ment of the intercalated organic molecules.

In 1970, R.C.Geballe and others found that the superconducting transition temperatures of both Ta- and Nb-dichalcogenides (MX_2) were strongly influenced by absorbing pyridine molecules between successive MX_2 layers.[11] This phenomenum was explained in consideration of charge transfer of the lone pair electron on the nitrogen atom of pyridine molecule to the MX_2 layer. Since the discovery that intercalation of organic molecules induce a large change in host material properties, organic-intercalation compounds have been spotlighted in the material sciences field and new types of intercalation compounds using a variety of host compounds as shown in Table 3. 1 have been synthesized and characterized.

Table 3. 1 Layer structure compounds found to intercalate.

Host Materials	Guest
Clay minerals	
Montmorillonite	amines, ammonia, hydradine,
Vermiculite	alcohols, glycols, proteins,
Kaolinite	aromatics
Ziconium phosphate	alcohols, amines,
	organometals, epoxides
Graphite	electron donor : alkali metals,
	alkaline earth metals
	elctron acceptor : halogens, halides,
	acidic oxides
Transition metal	
Dichalcogenides	alkali metals, amines, phosphines,
	organometals
Metal chalcophosphides	alkali metals, amines, organometals
Metal oxyhalides	alkali metals, amines, organometals
Miscellaneous	
MoO_3	alkali metals,
$H_2Ti_3O_7$	alkylamines
$Mg(Al)(OH)_2$	ClO_4^{2-}, CO_3^{2-}
$Ni(CN)_2$	pyridine
Ag_2C_2O	pyridine

3. 2 Reaction Mechanisms in Organic-Intercalation Compounds

Organic-intercalation compounds are classified into several groups on the basis of reaction mechanisms in the interlayer as shown in Fig. 3. 1.

(a) Storage of Organic Molecules in the Host Compound Interlayer

Organic molecules with OH- or NH_2-groups are easily intercalated to form OH-O or NH-O hydrogen bonding between the guest molecule and the lamelar surface of the host compound. Intercalated polar molecules such as acrylonitrile are also stabilized by forming

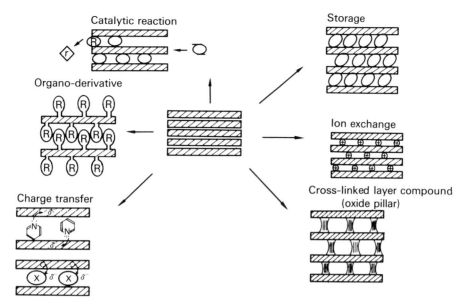

Fig. 3. 1 Reaction mechanisms in intercalation compounds.

a dipole-ion interaction between the guest molecules and the interlayer cations(e.g. exchangeable cation in clay mineral). These intercalation compounds are stable at a low temperature range but become unstable at high temperature, resulting in desorption of the intercalated molecules. This phenomenum was successfully applied to injection molding of epoxy polymer.[12,13] Montmorillonite was used as a host material to store amine molecules which act as catalysts to promote polymerization of epoxy monomers. The intercalated amine molecules are stably retained in the montmorillonite interlayer below 170°C. Therefore, amine molecules do not act as catalysts for polymerization of epoxy monomers during injection of the mixture of epoxy monomers and amine-montmorillonite complexes (injection temperature <150°C). In the temperature range from 180 to 200°C (polymerization temperature), the intercalated amine molecules are deintercalated to react with epoxy monomers surrounding montmorillonite, resulting in rapid polymerization into epoxy resin.

Intercalation of organic molecules plays an important role in heat-resistance and radiation-resistance of the organics, because the host material protects the intercalated organics against physical and chemical actions.

On the other hand, the ion-exchange ability of intercalation compounds is utilized for fixation of some harmful ions and for selective absorption of specified ions. Hydrated alkaline-ions or alkaline earth-ions in the interlayer of layered inorganic substances can be easily replaced with other cations in aqueous electrolyte solutions by ion-exchange reaction. An example of this kind of reaction is the exchange of alkaline ions in montmorillonite with harmful ions such as Cd and Cs ions in waste solutions, followed by heat treatment of the Cd or Cs-bearing montmorillonite to be convert into stable oxides. The titanium bronz bearing proton as interlayer ions, prepared by replacing the exchangeable Na ion in

NaTi$_3$O$_7$ with the proton, exhibited high selective absorption for lithium ion compared with other alkaline ions.[14] Many investigations have also been done on selective absorption of D- or L-optical isomers of organic molecules into layered inorganic substances.[15-17]

(b) Organic Derivatives of Layered Inorganic Compounds

Intercalation compounds are divided into three classes according to bonding characteristics between host layer and guest molecules. One is the absorption-type complex in which organic molecules are loosely bound to host layers by hydrogen bonds, ion-dipole interaction, charge-transfer bonds, etc. A second type of the host-guest interaction is seen in the ionic complex formed by replacing interlayer cations, e.g. alkaline or alkaline earth ions in clay minerals, with organic ions such as alkylammonium ions. The electrostatic force of attraction between host and guest plays an important role in stabilizing the intercalated organics. Intercalation compounds with the third type of bonding are organic derivatives of layered inorganic compounds. Organic molecules in the complexes are directly bonded to the lamellar surface by covalent bonds or by replacing the outermost atoms of the lamellar surface with organics.[13-20] The products are quite exotic compounds which have been given much attention in heterogeneous catalystic reactions.[18]

(c) Stereospecific Polymerization in the Interlayer

The third type of intercalation mechanism is stereospecific polymerization in the interlayer region which is considered to be a reaction vessel with pseudo two-dimensional space on the atomic scale. Protein- and amino acid-clay complexes have been studied from the viewpoint of the original synthesis of proteins from amino acids on the templates provided by the presence of clay particles in the primeval lakes. Organic molecules intercalated into host layers are restricted in their movement and are expected to form stereoregular polymers by irradiation or catalystic reaction. The first attempt to prepare two-dimensional macro molecules in the space between the silicate layers of clay minerals was made by Blumstein,[21] followed by many investigations on organic reactions in the interlayer of host materials.[22,23] Another example of stereospecific reaction is to synthesize coordination polymers with chain structures, using the lamellar surface of layered substances as a template. Characteristics of the Ni-rubeanic acid complex prepared in the interlayer of MoO$_3$ are described later.

(d) Charge Transfer between Guest and Host

Alkali and alkaline earth metals with low ionization potential react with the metal dichalcogenides with layered structure, graphite, etc., to form charge-transfer type intercalation compounds, in which the guest atoms are ionized to transfer a part of the electrons into the host layers. Drastic changes in the physical properties of the host substances are brought about by the charge transfer. The charge transfer from the guest to the host corresponds to reduction of the host substance, and the reverse charge transfer to oxidation of the host substance. Many studies on the reversible reaction between intercalation and deintercalation of alkaline atoms, which takes place electrochemically, have been focused on application to cathodic materials for secondary batteries.[24-26]

On the other hand, a large number of organic molecules with large pK_a values form the charge transfer type intercalation compounds with a variety of layered inorganic substances. One example is metal dichalcogenide-organic intercalation compounds which have been studied to clarify the relationship between charge transfer and the superconducting transition temperature of the parent metal dichalcogenides.[27-30]

Transition metal-oxyhalides and chalcogenophosphates also react with pyridine, alkylamine, etc., to form charge-transfer type intercalation compounds, resulting in remarkable changes in physical properties of the host substances. Details of preparation and characterization of these intercalation compounds are described in a later section.

3. 3 Stereospecific Polymerization in the Interlayer

The term "coordination polymer" has been used to describe a macromolecule produced by the reaction of a polydentate organic ligand with metal atoms. Such polymers have been widely studied, because many of them exhibit interesting properties in semiconductivity, magnetic susceptibility, catalytic activity and heat resistance.[31,32] However, neither molecular weight nor molecular structure is easily determined because of their low solubility and low degree of crystallinity.

Coordination polymers prepared by a conventional precipitation method have a complicated structure consisting of a linear chain structure and three-dimensional structure as shown in Fig. 3. 2, and their chemical composition depends largely on precipitation conditions. This structural ambiguity restricts the characterization of coordination compounds. In order to resolve this problem, stereospecific synthesis of coordination polymers

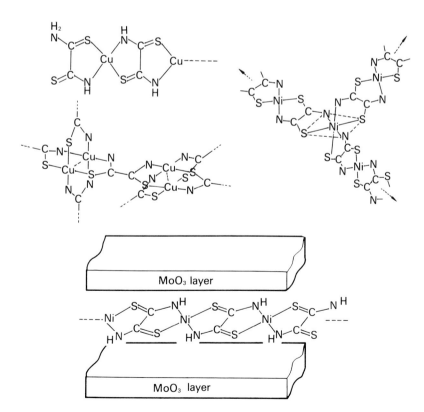

Fig. 3. 2 Schematic structural models of coordination polymer.

has been made by applying a monolayer method and by utilizing the two-dimensional interlayer space of layered inorganic substances.

Rubianic acid (RA) is a representative analytical regent with a tetradentate and reacts with various transition metals to produce coordination polymers which exhibit semiconductivity and antiferromagnetism. The Cu(11)-exchanged form of montmorillonite was used for the host material to intercalate RA molecules. However, only the monomeric complex, $Cu(RA)_2$, was formed in this case, because Cu(11) ion concentration in the interlayer was too low to bridge the distance between the neighboring Cu(11) ions with RA molecules.[33] In order to form the coordination polymer, M(RA), in the interlayer of host material, it is necessary to use another suitable layered substance as a host material in which the amounts of exchangeable cation can be controlled in a wide composition range. With a layered structure MoO_3 can intercalate hydrated sodium ions which are easily exchanged by various kinds of cations.[34] In an aqueous solution of $1N-Na_2S_2O_4$, MoO_3 was suspended at room temperature to form $Na_{0.5}(H_2O)_nMoO_3$. Then, the intercalated sodium ions were replaced by Ni(11) ions by treatment with a 1N-aqueous solution of $NiCl_2$ at 100°C for 1 day. The product with a chemical composition of $Ni_{0.25}(H_2O)_nMoO_3$ was soaked in an acetone solution saturated with RA at 60°C for a few days, washed several times with aceton to remove free RA molecules, and dried in vacuo. The basal spacing of 9.5Å was observed for the Ni(RA) intercalation MoO_3. A one dimensional electron density map of the Ni(RA)-MoO_3 complex, which was projected on the a-axis perpendicular to the MoO_3 sheet, indicated only one peak in the interlayer region of the host. Both this result and its interlayer space (4.6Å) suggest that the Ni(RA) complex formed in the interlayer of MoO_3 lies in such a way that the molecular plane of Ni(RA) complex is parallel to the MoO_3 layer.[35]

The infrared spectra of RA, $Ni_{0.25}(H_2O)_nMoO_3$ and Ni(RA)MoO_3 are shown in Fig. 3. 3. In the RA spectrum, three strong bands appear in the region of 3,000–3,500 cm^{-1}, which can be attributed to streching vibration of the NH_2 group, whereas the three bands are

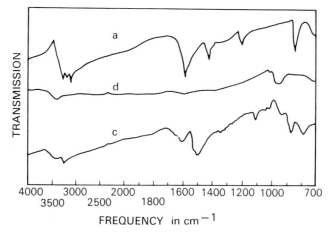

Fig. 3. 3 Infrared spectra (a) RA, (b) $Ni_{0.25}(H_2O)_nMoO_3$ and (c) Ni(RA)MoO_3.[35]

reduced to one near 3,250 cm^{-1} in the IR spectrum of the Ni(RA)-MoO$_3$ complex. A great spectrum change in the range of 1,800 to 700 cm^{-1}, related to the deformation and rocking modes of N-H, C-N and S-C-N groups, seems to be attributable to the chemical bond between the nitrogen or the sulfur of RA molecule and Ni(11) ions, because the coordination bond causes the frequencies of these groups to undergo considerable displacement.[35]

The Ni(RA) complex prepared by the usual precipitation method exhibited paramagnetism because the Ni(RA) complex consists of both coplanar (t_{2g}^6, a_{1g}^2, b_{1g}^0) and octahedral (t_{2g}^6, e_g^1, e_g^1) configurations of Ni(11) ions.[35] In contrast, the Ni(RA)-MoO$_3$ complex was nearly diamagnetic, while Ni$_{0.25}$(H$_2$O)$_n$MoO$_3$ exhibited paramagnetic behavior. These findings indicate that all Ni(11) ions in the Ni(RA) complex formed in the MoO$_3$ interlayer are coordinated with the nitrogen and the sulfur of the RA molecule to form a squared planar configuration.

Electrical resistivities of Na$_{0.5}$(H$_2$O)$_n$MoO$_3$ and Ni$_{0.25}$(H$_2$O)$_n$MoO$_3$ were in the 20-30 $\Omega \cdot$cm range at room temperature. The low electrical resistivities of Na- or Ni-intercalated MoO$_3$ originated in a charge transfer from the intercalated metal atoms to the MoO$_3$ layers, while the electrical registivity of the Ni(RA)MoO$_3$ complex was about 10^6 $\Omega \cdot$cm at room temperature. This high electrical resistivity could be explained by localization of a negative charge on the MoO$_3$ layer, caused by bond formation between the O atoms of MoO$_3$ layers and H$^+$ ions released from the N atoms of RA molecules through chelation with Ni(11) ions. The existence of an O-H bond was recognized by the fact that a broad band around 3,380 cm^{-1} attributable to stretching vibration of the O-H bond was found in the IR spectrum of the Ni(RA)MoO$_3$ complex.[35]

The Ni(RA) coordination polymer prepared in the MoO$_3$ interlayer was easily separated by dissolving only MoO$_3$ into boiling 1N-NH$_4$Cl solution for 1 hour.[37] The colloidal precipitates of Ni(RA) complex was suspended in water and centrifuged, followed by drying in vacuo. The IR spectrum of the Ni(RA) complex shows only one absorption band around 870 cm^{-1} assigned to stretching vibration of the C-S group in the RA molecule, although the Ni(RA) complex prepared by precipitation has two absorption bands at 872 and 835 cm^{-1}, indicating two different configurations around the Ni(11) ions, e. g. a square-planar configuration and an octahedral one.

Table 3. 2 Magnetic susceptibility of Ni-rubianic acid complexes; (a) Ni(RA)MoO$_3$, (b) Ni(RA) separated from the sample (a) by dissolving MoO$_3$, (c) Ni(RA) prepared by precitation method.
[after S. Son et al.[35] and S. Son et al.[37]]

Temperature (K)	104	205	273
Ni(RA)MoO$_3$	-0.8×10^{-6}	-2.0×10^{-6}	-1.3×10^{-6}
Ni(RA) (from IC comp.)	-1.1×10^{-7}	-1.9×10^{-6}	-1.1×10^{-6}
Ni(RA) (precip.)	13.2×10^{-6}	4.7×10^{-6}	4.1×10^{-6}

The Ni(RA) complex, which was prepared in the MoO_3 layer and then separated from the host, exhibits diamagnetism as shown in Table 3. 2. The diamagnetism of the complex indicates a squarplanar coordination about the Ni(11) ions.

From these results, the possibility of preparing stereospecific polymers using the pseudo two-dimensional interlayer space of host materials has been established.

3. 4 Charge Transfer Type Intercalation Compounds

As mentioned in a previous section, Geballe et al. found that the superconducting transition temperature (T_c) of the metal dichalcogenide (MX_2) was strongly influenced by intercalating pyridine molecules between the successive MX_2 layers.[11] A successive study on a homologous series of n-alkylamine intercalation TaS_2 also produced interesting results: the magnitude of the c-axis and the T_c of the complexes were dependent on the number of carbons (n) in alkylamine, $CH_3(CH_2)_nNH_2$.[38] As shown in Fig. 3. 4, two regions can be distinguished in the plots of both the interlayer spacing and T_c vs. the number of carbons in alkylamine. In Region 1 ($n = 1-4$), the interlayer spacing is independent of the n-value, indicating that the alkyl chain is arrangel parallel to the TaS_2 layer. The T_c values decrease with increasing chain length, while in the range of $n>8$, the interlayer spacing increases linearly with n. The slope, $\Delta d/\Delta n = 2.5$Å, can be explained by a β-type configuration, in which the alkylamine chain axis tilts by about 90° against the TaS_2 layer. The T_c keeps constant, independent of n in this region. Corresponding to the configuration of alkylamine molecules, the numbers of $-NH_2$ per unit area of the TaS_2 layer decrease with increasing n in Region 1, but do not change in Region 2. These results strongly suggest that the dramatic change of T_c was induced by the partial charge transfer from the alkylamine molecule N atom to the TaS_2 layer.

Bray and Sauer conducted studies on intercalation compounds, TaS_2(pyridine derivative)$_{1/n}$ applying the N^{14}-NQR measurement to clarify the charge transfer between

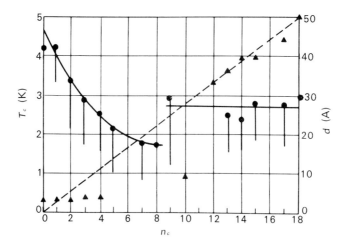

Fig. 3. 4 Increase of interlayer spacing (d) and onset temperature of superconductivity (T_c) plotted against the number of C atoms (n_c) for the TaS_2 (n-alkylamine) compounds. [after F. R. Gamble et al.[38]]

TaS$_2$ and guest molecules.[29] They found a linear relationship between T_c of the intercalation compounds and a charge density parameter of the pure guest molecules, $(l-\sigma_{NC})$, where l is the lone-pair electron density on the N atom of the pyridine ring and σ_{NC} is the charge density on the N–C bond. No correlation was observed between T_c and $(\pi-\sigma_{NC})$. Though the NQR measurement of the pyridine-derivatives intercalated in the interlayer of TaS$_2$ could give more precise conclusions for the charge transfer mechanism, their results indicate that the charge transfer from the N atom to the host layer plays an important role in changing the T_c of TaS$_2$.

Alkaline metals react to form intercalation compounds with semiconductive MX$_2$ in the same way with metallic MX$_2$ as host substances. Among these reactions, the intercalation of Na atoms into MoS$_2$ is the most representative one, in which electric properties of the host substance change from insulating to metallic behavior[39] (superconduction at low temperature). While the organic intercalation MX$_2$ complexes were prepared only when host materials have a partially filled conduction band. Therefore, it has been said that metallic compounds with layered structure such as TaS$_2$ can form the charge transfer-type organic intercalation compounds, but semiconductors such as MoS$_2$ are inert for organic intercalation.

Hargenmüller et al. studied the intercalation reaction between metal oxyhalides and amine molecules.[19,20] They succeeded in synthesizing many kinds of intercalation compounds such as AlOCl(NH$_3$)$_{1/3}$ (absorption-type complex) and AlONH$_2$ (derivative-type complex). FeOCl reacts with pyridine to form a charge transfer-type intercalation compound, resulting in remarkable changes in electrical and magnetic FeOCl properties.[40] FeOCl belongs to the orthorhombic space group P_{mnm} with $a = 3.780$, $b = 7.917$, $c = 3.302$Å and $z = 2$.[41] The crystal structure consists of a stack of double layered sheets as shown in Fig. 3.5(a). This structure is analogous to that of γ-FeOOH. The outermost atoms on each of the FeOCl layers are the Cl$^-$ ions. The interlayer Cl$^-$–Cl$^-$ distance, 3.680Å, is approximately twice the van der Waals radius of Cl. The reaction with pyridine was conducted in the closed system using distilled pyridine. The intercalation complexes prepared were black in color, and had two selected compositions depending on reaction temperature,[42] i. e. FeOCl(pyridine)$_{1/3}$ above 60°C and FeOCl(pyridine)$_{1/4}$ below 60°C. The kinetics of this intercalation were studied by Kikkawa.[43] He indicated that the reaction mechanisms were understood through nucleation and diffusion processes. In a lower temperature region, nucleation rate is represented by the first-order rate equation, while the two-dimensional Avrami-Ercfeev equation clearly explains the nucleation process in the higher temperature region,

$$-\ln(1-\beta) = k't \quad (T<60°C)$$
$$[-\ln(1-\alpha)]^{1/2} = k''t \quad (T>60°C)$$

were α and β are fractional intercalations for the final products of FeOCl(pyridine)$_{1/3}$ and FeOCl(pyridine)$_{1/4}$, respectively. The diffusion process in both temperature regions obeyed the two dimensional Jander's diffusion equation. Activation energies of all processes were about 10 kcal/mol.

A thermochemical study on FeOCl(pyridine)$_{1/3}$ was done using a calorimeter, and

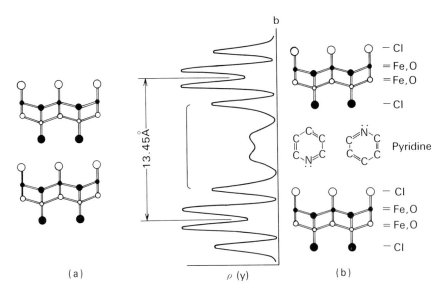

Fig. 3. 5 Schematic crystal structure of (a) FeOCl and (b) FeOCl(py)$_{1/3}$.[42]

enthalpy for the intercalation was found to be 5 kcal/mol.[44]

Both of FeOCl(pyridine)$_{1/3}$ and FeOCl(pyridine)$_{1/4}$ had the same orthorhombic unit cell with $a = 3.78$, $b = 13.45$, $c = 3.30$ Å, and $z = 2$.[42] A comparison of the lattice constants between FeOCl and its pyridine-intercalation complexes indicates that intercalation of pyridine molecules produces a small effect on atomic arrangements in the layered sheet of FeOCl, but only on interlayer separation.

One dimensional electron density projection on the b-axis shows two broad peaks in the interlayer region between adjacent FeOCl layers, indicating that the intercalated pyridine molecules are arranged with the molecular plane perpendicular to the FeOCl layer as shown in Fig. 3. 5(b).

Mössbauer effects on the oriented specimen of the FeOCl(pyridine)$_{1/3}$ complex such that the b-plane perpendicular to the incident γ-ray beam, were measured in the temperature range between 4.2 and 298 K using the γ-ray source Co^{57} doped in Cu metal kept at room temperature.[40] The Mössbauer spectrum at room temperature consists of a quadrupole couple with the measured values of 0.36 mm/s for the isomer shift (IS) and 0.92 mm/s for quadrupole splitting (ΔQS). The intensity ratio of the higher energy absorption line to the lower one is approximately 5/3, which is expected when the principal axis of the electric field gradient (EFG) is perpendicular to the incident beam and V_{zz} is negative. The hyperfine spectrum is observed at a temperature range below 60 K. The spectrum at 4. 2 K consists of only one set of a six-line spectrum whose relative intensity ratio is nearly 3 : 4 : 1, indicating that the internal magnetic field is oriented perpendicular to the b-axis. The internal magnetic field is along the z-axis of the EFG tensor because $\Delta s = s_1 - s_2$ is twice ΔQS in a paramagnetic state. The magnetic structure of FeOCl is rather complicated;[45] the direction of electron spin of half iron ions is parallel to the b-axis and

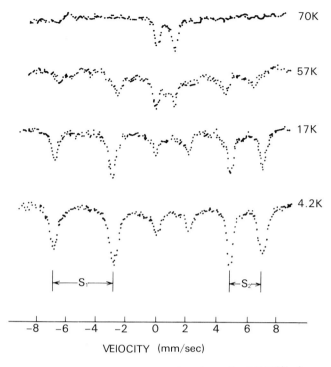

Fig. 3. 6 Mössbauer spectra of oriented sample of FeOCl(py)$_{1/3}$ (γ-ray$//b$).[40]

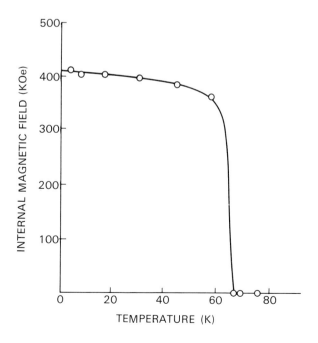

Fig. 3. 7 Temperature dependence of the internal magnetic field of FeOCl(py)$_{1/3}$.[40]

that of the other half is parallel to the c-axis, while the electron spin of whole iron ions in the FeOCl(pyridine)$_{1/3}$ complex is parallel to the a-axis. As seen in Fig. 3. 7, the temperature dependence of the internal magnetic field does not follow a Brillouin function, i. e. H_{eff} decreases slowly with rise in temperature up to 60 K, but above 60 K abruptly. The magnetic ordering temperature is estimated to be 65 K.

The measurements of electrical resistivity along the c-axis of FeOCl and FeOCl(pyridine)$_{1/3}$ were carried out in the temperature range between 200 and 373 K by a standard four-probe technique.[46] The value of FeOCl(pyridine)$_{1/3}$ is 1.2 Ω·cm at room temperature, being smaller by a factor of 10^7 than that of FeOCl. The remarkable changes not only in the magnetic property but also in the electrical conductivity of FeOCl were induced by intercalation of the pyridine molecule. Those changes are well explained by considering the partial electron transfer of the lone pair electron on the N atom of the pyridine molecule to the FeOCl layer. The charge transfer model is also in agreement with the ESR spectra of FeOCl(pyridene)$_{1/3}$, in which a broad siglet peak with $H_{msl} = 200$ gauss was observed at $g = 2.003$, though no spectrum was detected from FeOCl.[42]

If a charge transfer mechanism is significant in the hostguest interaction in the organic intercalation FeOCl, the pK_a value of the guest molecules would be qualitatively related to the change in FeOCl properties induced by intercalation of organic molecules. Pyridine derivatives and n-propylamine were used as guest molecules having different pK_a values (Lewis basicity). Since dimethylpyridine (DMPy), trimethylpyridine (TMPy) and n-propylamine (PA) are liquid at room temperature, FeOCl was directly soaked into these liquids. The aminopyridine (APy) acetone solution was used for solid APy. The black intercalation compounds were obtained after reaction with these organic reagents for about one week. The interlayer distance between FeOCl layers are expanded by intercalation of the guest molecules as represented in Table 3. 3. The chemical compositions of the intercalation compounds were determined by both thermogravimetry and chemical analysis of the compounds. In FeOCl(APy)$_{1/4}$ and FeOCl(DMPy)$_{1/4}$ compounds, the intercalated organic

Table 3.3 Interlayer distance (b) and electrical conductivity of FeOCl and FeOCl (organics)$_{1/n}$.
[after S. Kikkawa et al.[48]]

	pK_a	$b(\text{Å})$	$\rho(\Omega\cdot\text{cm})$	$E_a(\text{eV})$
FeOCl		7.92	10^6	0.6
FeOCl-(Py)$_{1/4}$	5.2	13.27	10	0.2
FeOCl-(DMP)$_{1/4}$	6.8	14.98	10^2	0.3
FeOCl-(AP)$_{1/4}$	9.2	13.57	10^3	0.2
FeOCl-(TMP)$_{1/6}$	9.6	11.79	10^3	0.2
FeOCl-(PA)$_{1/4}$	10.5	11.89	10^2	0.2

molecules are placed between the FeOCl layers so that the plane of the pyridine ring is almost perpendicular to the FeOCl layer, while the guest molecules in FeOCl(TMPy)$_{1/6}$ and FeOCl(PA)$_{1/4}$ compounds are arranged in such a way that the pyridine ring of TMPy and molecular axis of PA are nearly parallel to the FeOCl layer, respectively.[47] The deintercalation temperature of the guest molecules rose in the order of FeOCl(Py)$_{1/4}$, FeOCl(DMPy)$_{1/4}$, FeOCl(AP)$_{1/4}$ and FeOCl(TMPy)$_{1/6}$, indicating that the interaction between Cl$^-$ ions on the surface of the FeOCl layer and $-NH_2$ or $-CH_3$ of the guest molecules played an important role in stabilizing the complexes. This assumption is supported by a remarkable shift in the NH band toward a lower wave number.[48] The intercalation caused a large increase in electrical conductivity which increased with decreasing pK_a, suggesting that charge carriers in the compounds tend to localize with increasing interaction between hosts and FeOCl.[49]

The Mössbauer parameters depend on pK_a values of the guest molecules. The IS values of the intercalation complexes are slightly larger than that of FeOCl, suggesting increased Fe ions d-electrons in the intercalated FeOCl complexes.[48]

Other metal oxyhalides, MOCl (M ; V, Cr, In), also react to form intercalation compounds with organics having large pK_a.[50]

As mentioned above, metallic MX_2 with layered structure such as TaS_2 can react with organic substances to form charge transfer type intercalation compounds, while semiconductive MX_2 such as MoS_2 are inert for organic intercalation, though alkaline atoms are intercalated in the interlayer space of MoS_2. However, FeOCl reacts with various organic molecules having a large pK_a value to form intercalation compounds, even though FeOCl is a semiconductive substance. This result suggests that some semiconductive layered substances can also form the charge transfer type intercalation complexes. However, as FeOCl is an ionic crystal, it would be useful to conduct a further study on this problem using semiconductive layered substances with strong covalent bonds.

The crystal structure of MPX_3 consists of a stack of MPX_3 layers,[51] in which both M and P_2-dimer are octahedrally coordinated by six X atoms and the interlayer force between adjacent layers is weak. Among of them, $MPSe_3$ (M ; Mn, Fe) has a CdI_2-type

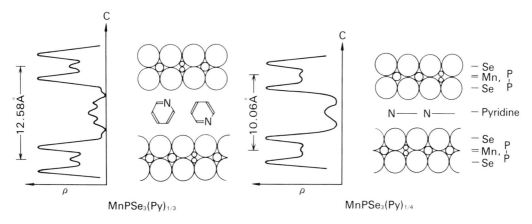

Fig. 3. 8 One dimensional electron density map and schematic representation of the structure of $MnPSe_3(Py)_{1/n}$ ($n = 3, 4$).[52]

structure similar to that of TaS₂ but is an insulator. In the MnPSe₃-pyridine system, MnPSe₃(pyridine)$_{1/3}$ was obtained by heating MnPSe₃ at 30°C for 4 days under saturated vapor pressure of pyridine. The intercalation compound was converted to another intercalation compound, MnPSe₃(pyridine)$_{1/4}$ by exposure to dry air for 1 day. The latter compound was more stable in air.[52] In both compounds, the a parameter was the same, but the c axis differed.

One-dimensional electron density projection on the c axis of MnPSe₃(pyridine)$_{1/3}$ indicates that the plane of the pyridine molecules perpendicular to the MnPSe₃ layer and the twofold axis of the pyridine ring form an angle to the layer such that the lone-pair electron cloud looks toward the MnPSe₃ layer. In the compound with the 10.1Å interlayer, the pyridine ring plane was parallel to the MnPSe₃ layer. This difference in configuration between two intercalation compounds was reflected in the electrical resistivity of these compounds. The resistivity of the host substance and MnPSe₃(pyridine)$_{1/4}$ at room temperature was higher than 10^8 Ω·cm, but the electrical conductivity of MnPSe₃(pyridine)$_{1/3}$ was larger by a factor of 10^3 than that of the host substance. The difference in electrical properties, which arose from the different orientation of the lone-pair electron's orbital of the N atom of pyridine molecule in the interlayer, suggest that semiconductive MnPSe₃ forms a charge transfer type intercalation compound.

In the case of FePSe₃, only one intercalation compound, FePSe₃(pyridine)$_{1/3}$, was obtained and the c-axis expanded as much as 5.9Å by intercalating pyridine molecules. The crystal structure of the FePSe₃(pyridine)$_{1/3}$ compound was thought to be similar to that of MnPSe₃(pyridine)$_{1/3}$. The electrical resistivity of the compound was about 10^6 Ω·cm at room temperature, the value of which was smaller by a factor of 10^2 than that of the host substance.[52]

MPS₃ (M ; Mg, Ca, Mn, Fe, Co, Ni, Zn etc.) which have a CdCl₂ type structure, react with alkylamines and organo-metallic compounds (Co(C₅H₅)₂ etc.) to form various kinds of intercalation compounds.[53,54]

As stated above, the reversible reaction between intercalation (reduction for host substance) and deintercalation (oxidation for host substance) in the system of an alkaline atom (especially the Li atom) and layered inorganic substance is spotlighted in the application of these compounds as electrode materials for a secondary battery. Many investigations have been done thus far on various kinds of layered inorganic substances (TiS₂,[55] FeOCl,[56] NiPS₃[57] etc.) to find suitable host materials with high energy density, excellent reaction reversibility, a wide composition range and little change in free energy over the composition range.

Among of them, TiS₂ is one of the most promising cathode materials, because TiS₂ exhibits good electronic conductivity and reacts with the Li atom to form a single non-stoichiometric phase with wide x values from 0 to 1 in Li$_x$TiS₂. The energy of Li-intercalation in TiS₂ is -206 kjmol^{-1} (480 Whkg^{-1}).[58] Recently, the effects of non-stoichiometry and solvents on the discharge property of the Li/TiS₂ battery were examined by Yamamoto et al.[25] Their conclusions were as follows: excess Ti reduces the utilization of the Li/TiS₂ battery, but was effective in pinning TiS₂ layers to each other to prevent the co-intercalation of solvents. Co-intercalation of the polar solvent also reduced the cell utilization. Tetrahydrofuran was the most favorable solvent because of difficulty for co-intercalation. However, open-circuit voltages of Li/TiS$_2$ gradually decreased from about 2.5 to 1.8 V with increasing x in Li$_x$TiS$_2$.[59] This phenomenum is an undesirable problem for practical use as a cathode material. Recently, metal trichalcogenides MX$_3$ (M; Ti, Zr, Nb, Ta, X; S, Se) with a layered type structure[60] have been noted as good cathode materials because the open-circuit voltage remains constant during discharge in the composition range of $0 \leq x \leq 2$ in Li$_x$MX$_3$.[26,61] The maximum Li contents per formular unit of host substances are 3Li for MX$_3$, the value of which is three times that of MX$_2$. Kikkawa et al. prepared TaSe$_3$, TaS$_3$ and NbS$_3$ with a crystal structure very similar to that of NbSe$_3$ by applying a high pressure technique (2GPa, 700°C).[62] Among these MX$_3$ substances, NbS$_3$ exhibited high open-circuit voltage and high constancy of OCE during discharge.

3.5 Organic Derivative of Layered Inorganic Substance

In reaction between MOCl (M; Al, Fe) and amine, Hargenmüllr et al.[19,20] succeeded in producing a new type of intercalation compound, in which intercalated NH$_3$ or CH$_3$NH$_2$ molecules are directly bonded to the host layer by replacing the outermost Cl atom of the MOCl layer with NH$_2$ or CH$_3$NH, resulting in formation of an M-N bond. The same type of compound was obtained by substitution of OCH$_3^-$ ion for Cl$^-$ ion in FeOCl.[63] FeOCl was reacted with an acetone solution of 4-aminopyridine at 110°C for 6 days in a sealed glass tube to form the intercalation compound, FeOCl(4-Apy)$_{1/4}$. This compound was then soaked in methanol in a sealed glass tube for at 80-110°C. Brown crystals with FeOOCH$_3$ chemical composition were produced.

$$FeOCl(Apy)_{1/4} + CH_3OH \rightarrow FeOOCH_3 + HCl + Apy$$

This reaction was accompanied by a remarkable change in interlayer distance between the host layers from 13.57 (FeOCl(Apy)$_{1/4}$) to 9.97Å (FeOOCH$_3$). A schematic representation of the FeOOCH$_3$ structure is derived on the base of the one-dimensional electron density map projected on the b axis and the IR spectra which indicates two kinds of CO bond orientation corresponding to splitting of the CO band.

Ethylene glycol (EG), OH(CH$_2$)$_2$OH, reacts with FeOCl(Apy)$_{1/4}$ to form a layered organic derivative with an FeO(O$_2$C$_2$H$_4$)$_{1/2}$ chemical composition. The interlayer distance of the compound was 10.89Å. Because the EG molecule has two OH groups at both ends of the alkyl chain, it is assumed that the EG molecule bridges two adjacent host layers. However, the one-dimensional electron density projection on the b axis of this compound indicates that the ionized EG molecule, O$_2$C$_2$H$_4^{2-}$, coordinates to two Fe ions in the same host layer as shown in Fig. 3.9(b). This intralayer bonding model is supported by the IR data (indicating only a gauche configuration) and the observed interlayer distance (obs: 10.89Å, interlayer bonding model: 8.2Å, intralayer bonding model: 11.0Å).

The Mössbauer parameters of FeOCl and its organic derivative are tabulated in Table 3.4. The IS values of the intercalation compound are almost the same as that of FeOCl, indicating that Fe ions remain in trivalent state in the intercalation compound. The QS values of both complexes are 0.59-0.6 mm/s being much smaller than 0.92 mm/s for FeOCl. The Fe ion in FeOCl is in an octahedral site which consists of four O^{2-} and two

Cl⁻ ions. This asymmetric charge distribution around the Fe ion caused large QS value for FeOCl. Replacement of Cl⁻ ion by OCH_3^- reduced the QS value as seen in Table 3.4 because crystal field symmetry around Fe ion in the organic derivatives is increased by forming an octahedral site with 6 oxygens ($4O^{2-}$ ions and 2 oxygen of organics).

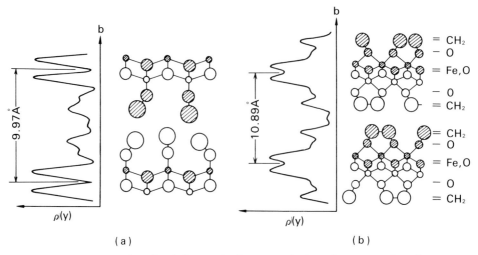

Fig. 3. 9 One dimensional electron density map and schematic representation of the structure of $FeOOCH_3$ and $FeO(O_2C_2H_4)_{1/2}$.[63,64]

Table 3. 4 Mössbauer parameters (mm/s) of FeOCl and its organic derivatives. [after S. Kikkawa et al.[64]]

	Isomer Shift (mm/s)	Quadrupole Splitting (mm/s)
FeOCl	0.36	0.92
$FeOCL(O_2C_2H_4)_{1/2}$	0.38	0.59
$FeOOCH_3$	0.37	0.60

No change was observed in direct reaction between FeOCl and CH_3OH or $C_2H_4(OH)_2$. Both $FeOOCH_3$ and $FeO(C_2H_4O_2)_{1/2}$ complexes were obtained using the $FeOCl(Apy)_{1/4}$ compound as a host material. This means that intercalation of CH_3OH or $C_2H_4(OH)_2$ molecules into the interlayer of FeOCl can take place by expanding the interlayer of FeOCl with intercalation of the Apy molecule, while, sodium methoxide, CH_3ONa, directly react with FeOCl to form $FeOOCH_3$.[9] The reaction between CH_3ONa and FeOCl was conducted at 60°C for 10 h and was accompanied by color change from reddish violet to brown. During this reaction no coloring in solution and no changes in external shape were recognized, indicating that the reaction had proceeded topotaxially, i.e. the Fe and O double layer remained after replacement of Cl⁻ by CH_3O^-. The results of the X-ray diffractogram, IR, DTA-TG and Mössbauer effect revealed that the product with interlayer distance of 10Å was the same as the compound obtained by reaction between $FeOCl(Apy)_{1/4}$ and CH_3OH.

VOCl has a crystal structure similar to that of FeOCl and intercalates pyridine molecules, but $VOOCH_3$ was not obtained. This difference in reaction with CH_3OH

between FeOCl and VOCl is an interesting problem for investigation in the formation of the graft-type intercalation compound.

Zirconium bis-monohydrogen orthophosphate, $Zr(HPO_4)_2 nH_2O$ has a layered structure (basal spacing: 12.3Å) in which reactive HPO_4 groups are situated on both sides of each layer.[65] There are two polymorphs of $Zr(HPO_4)_2 nH_2O$, i. e. α- and γ-$Zr(HPO_4)_2 nH_2O$. An organic derivative with a basal spacing of 18.4Å is formed in the reaction of γ-$Zr(HPO_4)_2 2H_2O$ with ethylene oxide.[66] The result of X-ray diffraction, chemical analysis and IR spectrum of the product indicated that the intercalated ethylene oxide molecule was subjected to react with the phosphate group of the host layer and to form a P-O-C ester bond as follows:

$$Zr(HPO_4)_2 + 2CH_2CH_2O \rightarrow Zr(HOCH_2CH_2OPO_3)_2$$

The structure of the compound is schematically represented in Fig. 3.10.[67] Recently, Yamanaka et al. found that the phosphate group of γ-$Zr(HPO_4)_2 H_2O$ can be topochemically replaced with various kinds of phospholic ester groups, $ROPO_3^{2-}$, by immersing γ-$Zr(HPO_4)_2 2H_2O$ in the solutions containing desired phospholic ester groups. Applying this phenomena to the synthesis of intercalated $Zr(HPO_4)_2$ compounds, they succeeded in preparing many organic derivatives of $Zr(HPO_4)_2$.[68-70]

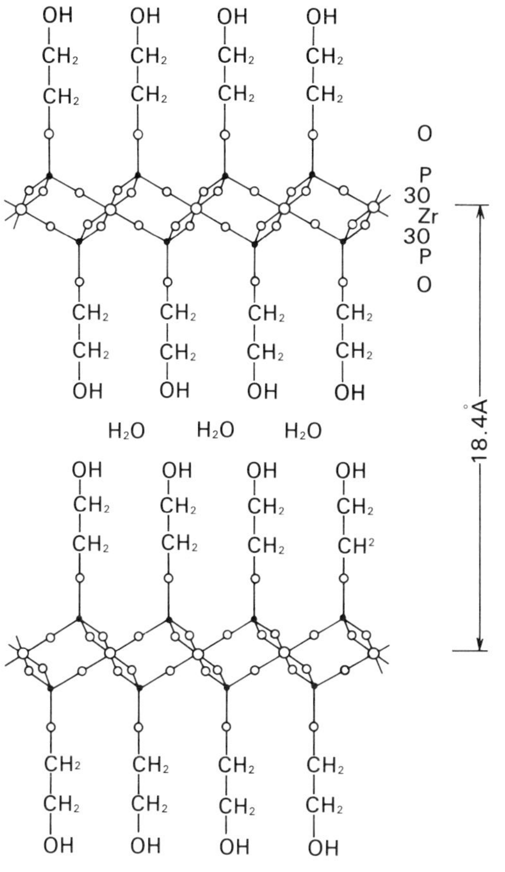

Fig. 3.10 Schematic representation of the structure of $Zr(HOCH_2CH_2OPO_3)_2 H_2O$.[67]

3.6 Concluding Remarks

A large number and a wide variety of organic-intercalation compounds have been investigated in the fields of basic science and in applications depending on reaction mechanisms in the interlayer of host substances. In the present paper, several examples have been briefly described on stereospecific formation of coordination polymers in the interlayer, large changes of physical properties in charge transfer type intercalation compounds and the preparation of organic derivatives of layered inorganic substances. In addition to these problems, Li-intercalation compounds are presently receiving much attention in applications as a cathode material for secondary batteries, and some layered compounds, such as close-linking montmorillonite, are also being considered as catalysts for organic reactions.

References

1) G.W.Brindley and S.Ray : "Complexes of Ca-montmorillonite with primary monohydric alcohols," Am. Mineral., **49** (1964) 106.

2) F.Kanamaru and V.Vand : "The crystal structure of a clay-organic complex of 6-amino hexanoic acid and vermiculite," Am. Mineral., **55** (1970) 1550.

3) P.G.Slade, M.I.Telleria and E.W.Radoslovich : "The structures of ornithine-vermiculite and 6-amino hexanoic acid-vermiculite," Clay & Clay Minerals, **24** (1976) 134.

4) J.M.Serratosa:"Infrared study of benzonitrile (C_6H_5CN)-montmorillonite complexes," Am. Mineral., **53** (1968) 1244.

5) S.Yamanaka, F.Kanamaru and M.Koizumi : "Studies on the orientation of acrylonitrile absorbed on interlamellar surfaces," J. Phys. Chem., **79** (1975) 1285.

6) S.Yamanaka, F.Kanamaru and M.Koizumi : "Role of interlayer cations in the formation of acrylonitrile-montomorillonite complexes," J. Phys. Chem., **78** (1974) 42.

7) R.H.Herber:"Mossbauer effect studies of intercalation compounds," Acc. Chem. Res., **15** (1982) 216.

8) D.M.Glementz, T.F.Pinnavaia and M.M.Mortand : "Stereochemistry of hydrated copper(11) ions on the interlamellar surfaces of layer silicates, — An electron spin resonance study," J. Phys. Chem., **77** (1973) 196.

9) S.Son, S.Kikkawa, F.Kanamaru and M.Koizumi : "Imediate formation of a layered compound, $FeOCH_3$ by a topochemical reaction," Igorg. Chem., **19** (1980) 262.

10) S.Nagai, S.Ohnishi, I.Nitta, A.Tsunashima, F.Kanamaru and M.Koizumi : "ESR study of Cu(11) ion complexes absorbed on interlamellar surfaces of montmorillonite," Chem. Phys. Lett., **26** (1974) 517.

11) F.R.Gamble, F.J.DiSalvo, R.A.Klemm and T.H.Geballe : "Super counductivity in layered structure organomerallic crystals," Science, **168** (1970) 568.

12) N.Adachi, M.Koizumi and F.Kanamaru : "New molding materials containing clay-organic complexes : properties and application," Kogyo-Zairyo, **25**, 3 (1977) 58.

13) N.Adachi, M.Adachi : "Molding materials containing clay-organic complexes : Epohard-3000 series," Kogyo-Zairyo, **25**, 6 (1977) 22.

14) H.Izawa, S.Kikkawa and M.Koizumi : "Ionexchange and dehydration of layered titanates, $Na_2Ti_3O_7$ and $K_2Ti_2O_9$," J. Phys. Chem., **86** (1982) 5023.

15) E.T.Degens, J.Matheja and T.A.Jackson : "Template catalysis : asymmetric polymerization of amino-acids on clay minerals," Nature, **227** (1970) 492.

16) T.A.Jackson : "Preferential polymerization and adsorption of L-optical isomers of amino acids relative to D-optical isomers on kaolinite templates," Chem. Geol., **7** (1971) 295.

17) G.W.Brindley and A.Tsunashima : "Clay-

organic studies ; XX. Montmorillonite complexes with dioxane, morpholine, and piperidine — Mechanisms of formation," Clay & Clay Minerals, **20** (1972) 233.

18) S.Yamanaka, K.Yamasaka and S.Hattori : J. Inclusion Phenomena (in press).

19) P.Hagenmüller, J.Portier, B.Barbe and P.Bouclier : "Vergleichende Untersuchung der Einwirkung von Ammoniak und von Aminer auf die Oxidchloride von Alumminium und Eisen," Z. Anorg. Allg. Chem., **355** (1967) 209.

20) P.Hagenmüller, J.Rouxel and J.Portier : "Un nonreau compose du fer trivalent : l'oxyamidure $FeO(NH_2)$," Compt. Rend., **254** (1962) 2000.

21) A.Blumstein : "Polymerization of absorbed monolayers; II. Thermal degradation of the inserted polymer," J. Polym. Sci., **A3** (1965) 2655.

22) R.Blumstein, A.Blumstein and T.H.Vanderspurt : "Polymerization of adsorbed monolayers; IV. The two-dimensional structure of insertion polymers," J. Colloid Interface Sci., **31** (1969) 236.

23) B.K.G.Theng : "Formation of two-dimensional organic polymers on a mineral surface," Nature, **228** (1970) 853.

24) M.S.Whittingham : "Chemistry of intercalation compounds ; Metal guest in chalcogenide hosts," Prog. Solid State Chem., **12** (1978) 41.

25) T.Yamamoto, S.Kikkawa and M.Koizumi : "Effect of nonstoichiometry and solvent on discharge property of Li/TiS_2 buttery," J. Electrochem. Soc., **131** (1984) 1343.

26) D.W.Murphy and F.A.Trumbore : "The chemistoy of TiS_3 and $NbSe_3$ cathodes," J. Electrochem. Soc., **123** (1976) 960.

27) F.R.Gamble, J.H.Osiechi and F.J.DiSalvo : "Some superconducting inercalation complexes of TaS_2 and substituted pyridines," J. Chem. Phys., **55** (1971) 3525.

28) P.J.Bray and E.G.Sauer : "N^{14} nuclear quadrupole resonance in compounds used as intercalates in superconduction complexes of TaS_2," Solid State Commun., **11** (1972) 1239.

29) R.Schöllhorn, H.D.Sagefka and T.Butz : "Ionic bonding model of the pyridine intercalation compounds of layered transition metal dichalcogenides," Mater. Res. Bull., **14** (1979) 369.

30) R.Shöllhorn : "Reversible topotactic redox reactions of solids by electron/ion transfer," Angew. Chem. Int. Ed. Engl., **19** (1980) 983.

31) D.C.Bradley : "Organometellic coordination polymers involving alkoxo and alkylamins bridges," Inorg. Macromol. Rev., **1** (1970) 141.

32) A.Berlin, A.I.Sherle : "Synthesis and properties of polymeric phthalocyanines and their metal derivatives," Inorg. Macromol. Rev., **2** (1971) 235.

33) S.Son, S.Ueda, F.Kanamaru and M.Koizumi : "Synthesis and characterization of a complex of rubeanic acid and copper(11) montmorillonite," J. Phys. Chem., **80** (1976) 1780.

34) R. Schöllhorn, R. Kuhlmann and J. O. Besenhard : "Topotactic redox reaction and ion exchange of layered MoO_3 bronzes," Mater. Res. Bull., **11** (1976) 83.

35) S.Son, F.Kanamaru and M.Koizumi : "Synthesis and characterization of the nickel(11)-rubeanic acid complex on interlamellar surface of molybdenum trioxide," Inorg. Chem., **18** (1979) 400.

36) L.Menabue, G.C.Pellacani and G.Peyronel : "Electronic and far ir spectra of some polymeric complexes of dithioxamide and its N,N'-disubstituted derivatives," Inorg. Nucl. Chem. Lett., **10** (1974) 187.

37) S.Son, N.Kinomura, F.Kanamaru and M. Koizumi : "Separation of nickel(11)-rubeanic acid complex prepared on interlamellar surfaces of molybdenum trioxide," J.C.S.Dalton (1980) 1029.

38) F.R.Gamble, J.H.Osiecki, M.Cais, R.Pisharody, F.J.DiSalvo and T.H.Geballe : "Intercalation complexes of Lewis bases and layered sulfides : A large class of new superconductrs," Science, **174** (1974) 493.

39) R.B.Somoano, V.Hadek, A.Rembaum : "Alkali metal intercalates of molybdenum disulfide," J. Chem. Phys., **58** (1977) 697.

40) F.Kanamaru, M.Shimada, M.Koizumi, M.Takano and T.Takada : "Mössbauer effect of FeOCl-pyridine complex," J. Solid State Chem., **7** (1973) 297.

41) M.D.Lind : "Refinement of the crystal structure of iron oxychloride," Acta Cryst., **B26** (1970) 1058.

42) F.Kanamaru, S.Yamanaka and M.Koizumi:

"Synthesis and some properties of a layer-type inorganic-organic complex of FeOCl and pyridine," Chem. Lett., 373 (1974).

43) S.Kikkawa : "Kinetic study on the system of FeOCl and pyridine," J. Solid State Chem., **31** (1980) 249.

44) S.Yamanaka, T.Nagashima and M.Tanaka : "Heat of formation of an intercalation complex between iron(III) oxychloride and pyridine," Thermochim. Acta, **14** (1977) 236.

45) R.W.Grant : "Magnetic structure of FeOCl," J. Appl. Phys., **42** (1971) 1619.

46) F.Kanamaru and M.Koizumi : "Electrical property of an intercalated compound, $FeOCl(C_5H_5N)_{1/3}$," Jpn. J. Appl. Phys., **13** (1974) 1319.

47) S.Kikkawa, F.Kanamaru and M.Koizumi : "Preparation and properties of FeOCl-pyridine derivative complexes and their reactivity with methyl alcohol," Reactivity Solids, **8** (1977) 725.

48) S.Kikkawa, F.Kanamaru and M.Koizumi : "Intercalation compounds $FeOCl(pyridine \ derivatives)_{1/n}$ and $FeOCl(n-plopylamine)_{1/4}$," Bull. Chem. Soc. Jpn., **52** (1979) 963.

49) S.Kikkawa, F.Kanamaru and M.Koizumi : "Preparation and properties of intercalation compounds $FeOCl(organic \ compounds)_{1/n}$," Physica, **105B** (1981) 249.

50) S.Kikkawa, F.Kanamaru and M.Koizumi : unpublished data.

51) B.Taylor, J.Steger, A.Wold and E.Rostiner: "Preparation and properties of iron phosphorus triselenide, $FePSe_3$," Inorg. Chem., **13** (1974) 2719.

52) S.Otani, M.Shimada, F.Kanamaru and M.Koizumi : "Preparation and characterization of $MPSe_3(py)_{1/n}$ complexes," Inorg. Chem., **19** (1980) 1249.

53) S.Yamanaka, H.Kobayashi and M.Tanaka : "New intercalated complexes of $MPSe_3(M = Mg,Zn,Mn)$ with n-alkylamines," Chem. Lett., 329 (1976).

54) Y.Mathey, R.Clement, C.Sourisseau and G.Lucazeau : "Vibrational study of layered MPX_3 compounds and of some intercalates with $Co(\eta_5-C_5H_5)_2^+$ or $Cr(\eta_6-C_6H_6)_2^+$," Inorg. Chem., **19** (1980) 2773.

55) M.S.Whittingham : "The role of ternary phases in cathode reactions," J. Electrochem. Soc., **123** (1976) 315.

56) P.Palvadeau, L.Coic, J.Rouxel, J.Portier : "The lithium and molecular intercalates of FeOCl," Mater. Res. Bull., **13** (1978) 221.

57) A.H.Thompson and M.S.Whittingham : "Transition metal phosphorus trisulfides as battery cathods," Mater. Res. Bull., **12** (1977) 741.

58) M.S.Whittingham : "Electrical energy storage and intercation chemistry," Science, **192** (1976) 1126.

59) M.S.Whittingham : Electro Materials and Processes for Energy Conversion and Storage, ed. by J.D.E.McIntyne, S.Srinivasan and F.G. Will, The Electro chemical Society, Princeton, New Jersey (1977).

60) S.Furuseth, L.Brattas and A.Kjekshus : "On the crystal structure of TiS_3, ZrS_3, $ZrSe_3$, $ZrTe_3$, HfS_3 and $HfSe_3$," Acta Chem. Scand., **A29** (1975) 623.

61) R.R.Chianelli and M.B.Dines : "Reaction of n-butyllithium with transition metal trichalcogenides," Inorg. Chem., **14** (1975) 2417.

62) S.Kikkawa, N.Ogawa, M.Koizumi and Y. Onuki : "High-pressure synthesis of TaS_3, NbS_3, $TaSe_3$ and $NbSe_3$ with $NbSe_3$-type crystal structure," J. Solid State Chem., **41** (1982) 315.

63) S.Kikkawa, F.Kanamaru and M.Koizumi : "Synthesis and some properties of $FeOOCH_3$ — A new layered compound," Inorg. Chem., **15** (1976) 2195.

64) S.Kikkawa, F.Kanamaru and M.Koizumi : "Preparation and properties of $FeO(O_2C_2H_4)_{1/2}$," Inorg. Chem., **19** (1980) 259.

65) A.Clearfield and G.D.Smith : "The crystallography and structure of γ-zirconium bis(monohydrogen orthophosphate) monohydrate," Inorg. Chem., **8** (1969) 431.

66) S.Yamanaka, F.Kanamaru and M.Koizumi: "Organic derivative of γ-ziroconium phosphate," Nature Phys. Sci., **246** (1973) 63.

67) S.Yamanaka : "Synthesis and characterization of the organic derivatives of zirconium phosphate," Inorg. Chem., **15** (1976) 2811.

68) S.Yamanaka, S.Yamasaka and M.Hattori : "Exchange of interlayer monohydrogen orthophosphate ions of zirconium bis (monohydrogen orthophosphate) dihydrate with 1- and 2-glycero phosphate ions," J. Inorg. Nucl.

Chem., **43** (1981) 1659.
69) S. Yamanaka, S. Sakamoto and M. Hattori : "Phase transition in the layer structure compounds $Zr(HPO_4)(n\text{-}C_nH_{2n+1}PO_4)$," J. Phys. Chem., **85** (1981) 1930.
70) S. Yamanaka, S. Sakamoto and M. Hattori : "Mechanism for the heterogeneous exchange of the interlayer phosphate groups of γ-zirconium phosphate with phenyl phosphate groups," J. Phys. Chem., **88** (1984) 2067.

4. Piezoelectric Ceramic Actuator

Sadayuki TAKAHASHI*

Abstract

A piezoelectric ceramic actuator with multilayer internal electrodes has been fabricated by a tape-casting method. It has 144 internal electrode layers maintaining gaps of about 100 μm. It operates in a piezoelectric stiffened mode.

Typical properties of the actuator using PZT family piezoelectric ceramics are: a strain of about 0.1 %, a clamping stress of about 3.5×10^7 N/m^2 and an electromechanical coupling factor k of 70 % with 100 V applied.

It was confirmed that light polarization can be controlled by applying stress to an optical fiber using this actuator.

Keywords : piezoelectricity, ceramics, actuator, multilayer, stack, electromechanical coupling, tape-casting, field induced strain.

4.1 Introduction

Piezoelectricity is characterized by the behavior of certain crystalline materials in developing electrical polarization proportional to the mechanical stress and in developing mechanical strain proportional to the applied electrical field. It was discovered by J. and P.Curie in 1880. Today, piezoelectricity is applied to important electronic devices. The applications may be divided into two categories. One is as a resonator and the other is as an electro-acoustic transformer. Recently, there has been keen interest in a piezoelectric actuator as a third application.[1]

Electromagnetic actuators have been studied for many years and it seems that this technology has reached a saturated state. Studies aimed at achieving a piezoelectric actuator have just begun. However, such an actuator is attractive because it is expected to have lower power consumption, less heat generation and be smaller in size, compared with an electromagnetic actuator.

Piezoelectric materials are divided into five groups according to their forms, i.e., single crystal, ceramic, organic high polymer, thin film and composite. As actuator materials, ceramics are preferable for their high piezoelectric constant, high mechanical stiffness and availability.

It seems that two kinds of piezoelectric ceramic actuators are now in practical use.

---

* Fundamental Research Laboratories, NEC Corporation, 4-1-1, Miyazaki, Miyamae, Kawasaki, Kanagawa 213, Japan.

One is a bimorph element, which is composed of two piezoelectric ceramic plates. It operates based upon a piezoelectric unstiffened effect. Its electromechanical energy conversion efficiency is not very high. The other is a piezoelectric ceramic plate stack, bonded together with an adhesive or spring loaded.[2] It operates by a piezoelectric stiffened effect, so that it has high energy conversion efficiency. However, it requires relatively high driving voltage, because it is difficult to achieve a single thin ceramic plate.

This report deals with a study on internal electrode multilayer piezoelectric ceramic stack actuators, fabricated by a tape-casting method.[3-5] The actuator can be driven by a low voltage application, maintaining high energy conversion efficiency.

4. 2 Stack and Bimorph Comparison

Piezoelectricity has a stiffened and unstiffened effect. With the stiffened effect, the ceramic material elongates or contracts along the polar axis as a result of applying an electric field parallel to the polar axis. On the other hand, it elongates or contracts perpendicular to the polar axis with the unstiffened effect. The stack element is driven by the stiffened effect and the bimorph element is driven by the unstiffened effect. First of all, merits and demerits of both actuator elements are studied by theoretical calculations.

Structures of the stack and bimorph element are shown in Fig. 4. 1. The stack is composed of many ceramic plates bonded together with an adhesive. Individual ceramic plate surfaces are metallized and electrically connected in parallel. When an electrical field is applied to the element, it elongates along the longitudinal direction. The bimorph element is composed of two ceramic plates. When the lower end is clamped mechanically, the upper free end moves in the direction shown by the arrow. Total input electric energy UE, converted mechanical energy UM, electromechanical coupling factor k^2, induced

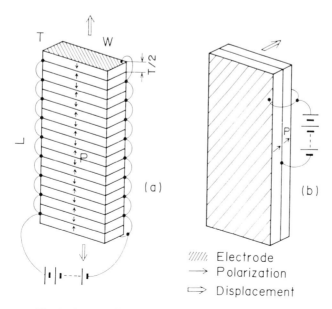

Fig. 4. 1 Two kinds of piezoelectric actuator elements.
(a) stack element, and (b) bimorph element.

displacement ξ, clamping force F and resonant frequency f were calculated for these two kinds of actuators according to the following formulas.[6]

$$UE_s = \frac{1}{2} C_s V^2 \tag{4.1}$$

$$UE_b = \frac{1}{2} C_b V^2 \tag{4.2}$$

$$UM_s = 2 \frac{LW}{T} \frac{d_{33}^2}{s_{33}^E} V^2 \tag{4.3}$$

$$UM_b = \frac{9}{8} \frac{LW}{T} \frac{d_{31}^2}{s_{11}^E} V^2 \tag{4.4}$$

$$k^2 = \frac{UM}{UE} \tag{4.5}$$

$$\xi_s = \frac{2L}{T} d_{33} V \tag{4.6}$$

$$\xi_b = 3 \left(\frac{L}{T}\right)^2 d_{31} V \tag{4.7}$$

$$F_s = 2W \frac{d_{33}}{s_{33}^E} V \tag{4.8}$$

$$F_b = \frac{3}{4} \frac{TW}{L} \frac{d_{31}}{s_{11}^E} V \tag{4.9}$$

$$f_s = \frac{1}{2L} \sqrt{\frac{1}{\rho s_{33}^D}} \tag{4.10}$$

$$f_b = \frac{1.875^2}{4\sqrt{3\pi}} \frac{T}{L^2} \sqrt{\frac{1}{\rho s_{11}^E}} \tag{4.11}$$

where, C: electrostatic capacitance,
V: applied voltage,
L, W, T: element size dimensions,
d_{31}, d_{33}: piezoelectric constant,
s_{11}, s_{33}: elastic compliance,
ρ: density,
s: suffix indicating stack element,

b: suffix indicating bimorph element.

Calculations were carried out for $V = 100$ V, $L = 20$ mm, $W = 5$ mm and $T = 0.5$ mm, using the constants for binary solid solution piezoelectric ceramics 0.65 Pb(Mg$_{1/3}$Nb$_{2/3}$)O$_3$–0.35 PbTiO$_3$. The constants are shown in Table 4.1.

Table 4.1 Various constants for binary solid solution ceramics 0.65Pb (Mg$_{1/3}$Nb$_{2/3}$) O$_3$–0.35PbTiO$_3$.

Electromechanical Coupling Factor	k_{33}	70%
	k_p	58
Piezoelectric Constant	d_{33}	5.63×10^{-10} m/V
	d_{31}	-2.41
Elastic Compliance	s_{33}^E	2.01×10^{-11} m^2/N
	s_{33}^D	1.03
	s_{11}^E	1.52
Dielectric Constant	ε_{33}^T	3,640

The results of the above calculations are shown in Table 4.2. The UE values for the stack and for bimorph are the same. However, the UM values differ. As a result, k_s^2 is about seven times larger than k_b^2. The F and f values for the stack are also larger than those for the bimorph. The bimorph is superior to the stack only in regard to induced displacement.

Table 4.2 Calculated actuator characteristics for stack and bimorph elements.

Element	UE ($\times 10^{-7}$J)	UM ($\times 10^{-7}$J)	k^2 (%)	ξ (μm)	F (N)	f (kHz)
Stack	1,290	630	49	4.5	28	88
Bimorph	1,290	86.2	6.7	116	0.15	0.59

4.3 Fabrication Process

The stack is an excellent actuator element. However, a conventional stack element requires high driving voltage and is unsuitable for miniaturization and mass production. These defects seem to prevent the stack being utilized as an electronic component. An internal electrode multilayer structure stack element has been studied as a way of overcoming these defects. This element is fabricated by a tape-casting method. The process flow chart is shown in Fig. 4.2. Piezoelectric ceramic calcined powder with some organic binder, plasticizer and solvent is milled into a slurry with balls. Poly-vinyl alcohol (PVA), butyl phthalyl butyl glycolate (BPBG) and ethyl alcohol are used as binder, plasticizer and solvent, respectively. The slurry is cast into a green sheet several tens of microns thick on a running organic film using a spatula vessel. After drying, the green sheet is stripped from the organic film and is cut into appropriate sizes. Metal paste is printed on it as the internal electrode. Platinum paste is usually used for this purpose. Silver paste is used if the piezoelectric ceramic can be sintered at temperatures as low as 1,000 °C. A number of green sheets with electrodes are laminated and pressed together, at temperatures as high as 100 °C.

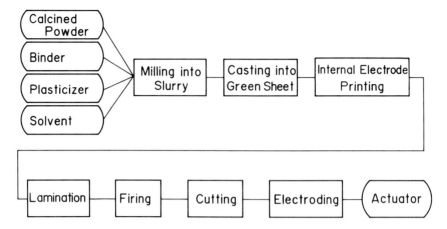

Fig. 4.2 Fabrication process flow chart for a multilayer piezoelectric ceramic actuator.

The pressed body is heated to about 500 °C for more than 100 h to break down the organic materials included in the slurry as binder, plasticizer or solvent. After the organic materials are broken down, the ceramic body is sintered in place at 1,000–1,300 °C. The sintered body is cut into appropriate sizes, according to the various projected uses.

Internal electrode layer ends appear on all four sides of the cut ceramic element. Insulator paste is strip coated onto alternate internal electrode layer ends and fired. The projecting ends of alternating electrode layers which have not been covered by the insulator strips, are still open to the air. These unprotected projecting ends extending from one face of the element are electrically connected together by a layer of conductive material coated over these ends. This forms a parallel pile of electrodes separated by ceramic layers, much like the pages of a book held together by the book binding back.

Next, these electrode layers ends projecting from the opposite face of the element are connected together in the same manner. (See Fig. 4.3.)

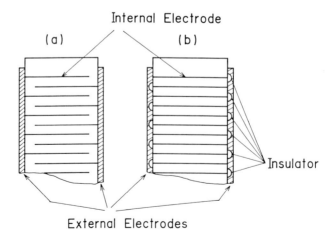

Fig. 4.3 Multilayer ceramic structure; (a) capacitor, and (b) actuator.

The thus obtained actuator element has a structure similar to that of a ceramic chip capacitor, but it does not have exactly the same structure, as shown in Fig. 4. 3. In the usual ceramic chip capacitor, the individual internal electrode area is smaller than the element cross section area. On the other hand, the actuator element has internal electrodes which extend to the full space area of the element cross section. If the actuator element has the same internal electrode structure as that of a ceramic chip capacitor, induced strain will become nonuniform and local internal stress will arise.

The local stress distribution mentioned above was estimated by the finite element method. A simplified element structure model used for finite element method analysis is shown in Fig. 4. 4 (a), where $H = 0.23$ mm, $lo = 1.5$ mm and $le = 1.0$ mm. The single electrode area is smaller than the element cross sectional area. The estimation was carried out using the constants listed in Table 4. 1 with 230 V applied across the electrodes. Induced stress distribution around the internal electrode end is shown in Fig. 4. 4 (b). It can be seen that stress is concentrated around the internal electrode end. Maximum tensile and compressive stress values are 1×10^8 and 1.2×10^8 N/m², respectively. This tensile stress is in the same order as the ceramics destruction stress. As a matter of fact, a stack element with internal electrodes of area smaller than the element cross sectional area, is destroyed mechanically by applying an electric field. The induced stress concentrated around the end of the internal electrode might cause the destruction. It was confirmed both theoretically

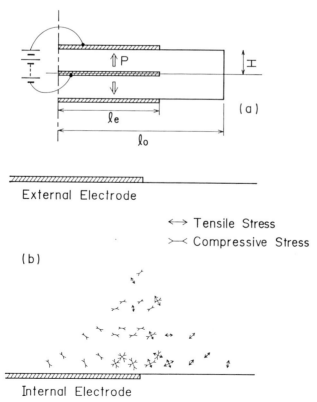

Fig. 4. 4 Finite element method analysis; (a) model, and (b) calculated stress distribution.

and experimentally that no stress arises when single electrodes are as large as the element cross section. Thus, a ceramic chip capacitor structure is unsuitable as an actuator element.

4.4 Actuator Characteristics

An internal electrode multilayer piezoelectric ceramic stack actuator, 18 mm in length with a 2 × 3 mm cross sectional area and 144 internal electrode layers maintaining gaps of 0.115 mm, was fabricated on an experimental basis using pseudo-ternary solid solution piezoelectric ceramics $(1-x-y)Pb(Ni_{1/3}Nb_{2/3})O_3-xPbTiO_3-yPbZrO_3$. This solid solution shows high field-induced strain and can be sintered at about 1,100 °C. A silver paladium metal alloy was used as the internal electrode material. Electrostatic capacitance for the element was 330 nF.

The displacement which arises in longitudinal direction versus applied voltage is shown in Fig. 4.5. Butterfly wing hysteresis, peculiar to ferroelectric ceramics, can be seen. In the initial state, the element does not elongate when a low voltage is applied, because spontaneous polarization is distributed homogeneously. By increasing the applied voltage to more than 50 V, the induced strain increases abruptly. The strain includes a piezoelectric strain and a strain which arises from the ferroelectric domain rearrangement. Even after the voltage is removed, residual strain can be seen. This material is now polarized. If, however, the voltage is reversed in polarity, the strain decreases to zero and again increases as the voltage increases. The field in which strain direction reverses is termed the coercive field. When the voltage is removed, residual strain can again be seen and the material is negatively polarized.

Displacement versus applied voltage for polarized material when voltage is applied

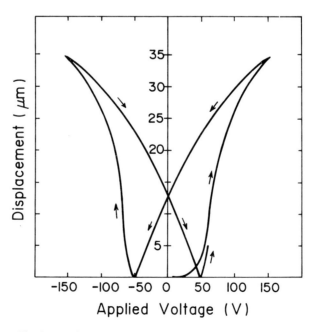

Fig. 4.5 Displacement versus applied voltage for an internal electrode actuator.

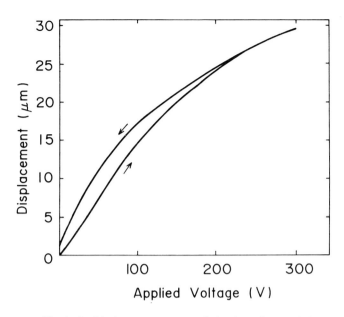

Fig. 4. 6 Displacement versus applied voltage for a polarized actuator.

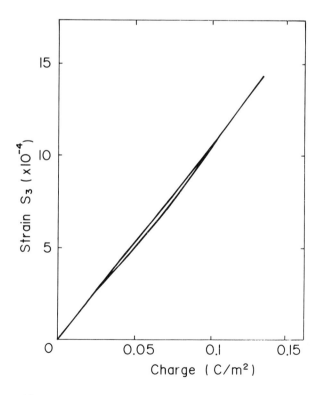

Fig. 4. 7 Strain S_3 parallel to the polarized direction versus electric charge stored up in the actuator.

in the same direction as the polarizing voltage is shown in Fig. 4. 6. Displacement does not increase linearly with the applied voltage. It is often desirable to track the displacement of an actuator in the open-loop mode. A monitoring field is not adequate because of the above nonlinear relationship. That length direction strain S_3, plotted against an electrical charge, reveals a linear relationship with virtually no hysteresis as shown in Fig. 4. 7.[7] By monitoring the charge, a very accurate displacement prediction can be made.

To estimate the response time for the induced strain, a rectangular voltage pulse with 100 V height and 0.4 msec length was applied to the element. The voltage across the element and the resultant displacement are shown in Figs. 4. 8 (a) and (b), respectively. The displacement is quickly induced and reduced in accordance with the voltage change. The response time is shorter than 10 μsec.

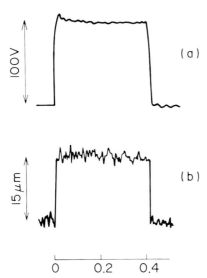

Fig. 4. 8 Displacement response to applied voltage: (a) applied voltage, and (b) generated displacement.

When a voltage is applied to the element, it will expand as a result of strain, if the element is free to move, with no accompanying stress generation. However, stress is generated under mechanical loading. Load-displacement characteristics are shown in Fig. 4. 9. Displacement represented by the Y-axis intercept is the displacement generated by applying 115 V without any load coupled to it. If the element is then subjected to force to restore its length to the original value, force represented by the X-axis intercept will be required.

The area of the triangle enclosed by the X- and Y-axes and the line which shows the load-displacement relationship, represents the mechanical energy which is stored in the element. On the other hand, electric input energy can be estimated according to Eq.(4. 1), listed previously. Hence, the electromechanical coupling factor k^2 can be determined using Eq.(4. 5). The value is 49 %.

Temperature sensitivity for displacement generated by applying 115 V is shown in Fig. 4. 10. The generated displacement changes about 10 % from 0 to 100 °C.

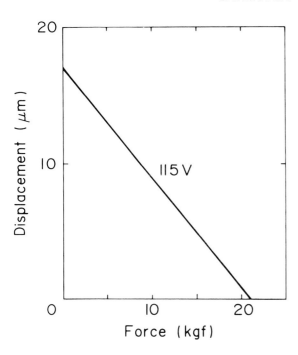

Fig. 4.9 Displacement versus force under a DC voltage of 115 V.

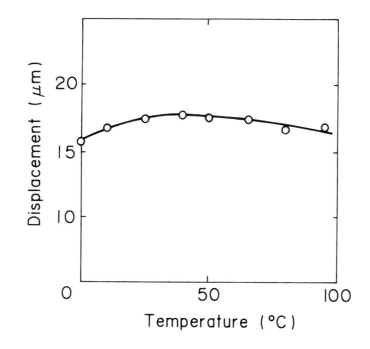

Fig. 4.10 Temperature dependence for a displacement, generated by a DC voltage of 115 V.

The element material is ferroelectric, so it includes DE hysteresis. If the element is operated continuously by applying an AC or pulsed DC voltage, its temperature will rise, due to the ferroelectric DE hysteresis loss. The element temperature will increase as the frequency and peak voltage increase. Element temperature versus elapsed time varying with the frequency of the applied field is shown in Fig. 4.11. Here, a rectified half sine voltage wave with a 100 V peak, was applied to the element. The temperature rises with time and frequency increase. However, it saturates in about 150 sec. Saturation may be caused by the balance between heat generation and heat radiation.

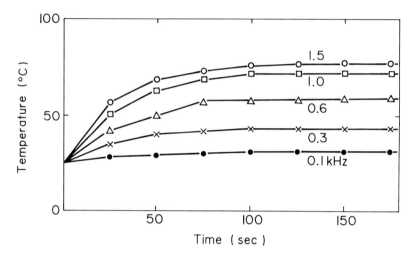

Fig. 4.11 Temperature of actuators when pulsed voltage is continuously applied.

4.5 Application

The multilayer piezoelectric ceramic stack actuator has extensive applications, such as displacement control for precision machining tools or optical instruments, a power generator for motors, relay or impact printer heads, high power electro-acoustic transformers etc. One application example, which the author is presently studying, is shown here.

In an optical heterodyne communication system, an optical information processing and optical sensor, it is very important to freely control light polarization. Although many methods have been proposed, the simplest method is to apply a force to a single mode optical fiber from opposite sides.

The experimental apparatus for light polarization control using a multilayer piezoelectric ceramic stack actuator is shown in Fig. 4.12. A single mode optical fiber of 125 μm in diameter with a spot size of 5.5 μm and a cut off wavelength 1.3 μm, was sandwiched by two piezoelectric actuators. A 1.55 μm wavelength semiconductor laser LD was used as a light source. The state of light polarization was measured while rotating a Rochon prism.

The relation between polarization P and applied voltage is shown in Fig. 4.13. Value P can be shown as a function of maximum light intensity I_{max} and minimum light intensity I_{min}.

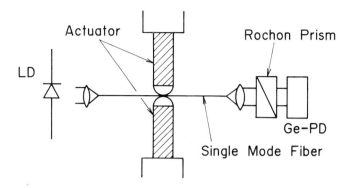

Fig. 4.12 Experimental construction for light polarization control using actuators.

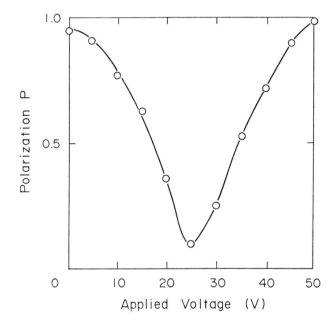

Fig. 4.13 Polarization degree P versus applied voltage.

$$P = \frac{I_{max} - I_{min}}{I_{max} + I_{min}} \tag{4.12}$$

States $P = 1$ and $P = 0$ correspond to linear polarization and circular polarization, respectively. As can be seen, a half wavelength voltage of about 50 V is obtained.

4.6 Summary

A multilayer piezoelectric ceramic actuator fabricated by a tape-casting method, and its application, are discussed. The actuator can be driven by a voltage of less than 100 V. Obtained free strain, parallel to the longitudinal direction, is about 0.1 % and clamping stress is about 3.5×10^7 N/m² in a 1 kV/mm field. Response time is less than 10 μsec. Generated strain is not linear to the applied voltage. However, it is linear to the electric charge which is stored in the actuator. The electromechanical coupling factor k^2 is about 50 %.

Light polarization control is easily realized by applying stress to a single mode optical fiber from both sides, using actuators. Half wavelength voltage is about 50 V.

Acknowledgements

The author wishes to thank Mr. Takeshi Yano, Mr. Tomezi Ohno, Mr. Hideo Takamizawa, Mr. Masatomo Yonezawa, Mr. Kazuaki Utsumi and Mr. Roh Ishikawa for their helpful discussion and encouragement.

References

1) K. Uchino : Essentials of Developments and Application of Piezoelectric Actuators, Electronic Essentials No. 3 (Nippon Kogyo Gijutsu Center, Tokyo) (1984) [in Japanese].
2) S. Yamashita : "Piezoelectric pile," Jpn. J. Appl. Phys. Suppl., **20** (1981) 93.
3) S. Takahashi, A. Ochi, M. Yonezawa, T. Yano, T. Hamatsuki and I. Fukui : "Internal electrode piezoelectric ceramic actuator," Ferroelectrics, **50** (1983) 181.
4) A. Ochi, M. Yonezawa and S. Takahashi : "Piezoelectric ceramic tiny actuators," Proc. 3rd Sensor Symp. (1983) 261.
5) S. Takahashi, A. Ochi, M. Yonezawa, T. Yano, T. Hamatsuki and I. Fukui : "Internal electrode piezoelectric ceramic actuator," Jpn. J. Appl. Phys. Suppl., **22** (1983) 157.
6) M. Toda and S. Osaka : "Electromotional device using PVF_2 multilayer bimorph," Trans. IECE Japan, **E61** (1978) 507.
7) C. Newcomb and I. Flinn : "Improving the linearity of piezoelectric ceramic actuators," Electron. Lett., **18** (1982) 442.

Original Review Papers

5. Toughening in ZrO_2-Based Materials

Katherine T. FABER[*1]

Abstract

Fracture toughening processes in transformation toughened zirconia and related systems are reviewed. The main toughening derives from the martensitic transformation of tetragonal ZrO_2 particles to their monoclinic form ahead of a crack as analyzed by thermodynamic and continuum approaches. The implications for crack-length dependent fracture resistance based upon the toughening models are included. Additional toughening contributions are evaluated for microcracking and crack deflection processes using recent models. Systems where transformation toughening occurs are summerized, including materials which are surface-strengthened through the tetragonal-monoclinic transformation. The effects of temperature and environment on mechanical behavior are also considered.

Keywords : toughness, transformation, microcracking, zirconia, R-curve, deflection.

5.1 Introduction

Over the past decade, a new family of ceramic materials has been developed which demonstrates unprecedented mechanical properties by inclusion of tetragonal zirconium oxide. In these materials, a crystallographic, martensitic transformation is stress-induced in a small zone ahead of a propagating crack which ultimately "shields" the crack from the applied loading. The stress-induced phenomenon has been of considerable interest for two main reasons: (a) the phenomenon provides increases in the fracture toughness which have been unsurpassed in any other monolithic ceramic, and (b) martensitic transformations, unknown in previously-studied ceramic systems, have received little or no attention by the ceramic community and provided a new avenue of research.

The first and classic example of transformation toughening was noted in partially stabilized zirconia (PSZ) containing MgO as a stabilizing agent.[*2,1)] The phenomenon has since been observed in a number of HfO_2- and ZrO_2-containing systems[2)] and has also been speculated to occur in enstatite, $MgSiO_3$.[3)] In addition to ZrO_2-based materials, other high modulus matrices have been used to incorporate ZrO_2 particles, including alumina, β''-

---

[*1] Assistant Professor, Department of Ceramic Engineering, The Ohio State University, 2041 College Road, Columbus, Ohio 43210, U.S.A.

[*2] Other stabilizers have also been utilized in transformation toughened systems, including CaO and rare-earth/oxides, such as Y_2O_3 and CeO_2.[2)]

alumina, and mullite, as well as non-oxide ceramic matrices, such as SiC and Si_3N_4.[4-8]

The fracture toughness increases in transformation-toughened materials range in magnitude from approximately 2 to 10 MPa m$^{1/2}$ over host materials (See Table 5. 1). As seen from the table, the magnitude of the toughness increase can be enhanced by choice of matrix. Yet, in each of these materials, the toughening increment clearly cannot be explained by considering the fracture toughness of each of the end members.

Table 5. 1 Fracture toughnesses of selected zirconia alloys.

Material	K_c (MPa m$^{1/2}$)	ΔK^T (MPa m$^{1/2}$)
Partially Stabilized Zirconia[4]	12.25	10.0
Tetragonal Zirconia Polycrystals[2]	>10	>8
Zirconia Toughened Al_2O_3[5]	8.5	4.5
Zirconia Toughened Mullite[2]	4.0	1.5
Zirconia Toughened Spinel[2]	4.15	2.0
Zirconia Toughened β''-Al_2O_3[6]	4.5	0.7-1.9
Zirconia Toughened SiC[7]	6.2	2.4
Zirconia Toughened Si_3N_4[8]	7.5	1.5

The focus of this paper will deal with mechanical aspects of transformation-toughened materials. (The crystallographic aspects and thermodynamics of the trnsformation may be found in References 9) and 10), respectively.) First, the mechanisms under which toughening is achieved will be reviewed, including the martensitic transformation and also the accompanying mechanisms of microcracking and crack deflection. R-curve effects, in which the resistance to fracture is related to the extent of crack growth, will be discussed in the light of the toughening mechanisms. A review of the environmental effects on the toughening, including high and low temperature aging and slow crack growth, will also be given.

5. 2 Fracture Toughening

The fracture toughening of the zirconia-based alloys is considered to primarily reside in the transformation phenomenon. However, at least two additional toughening mechanisms may be further enhance the fracture toughness: stress-induced microcracking and crack deflection. Their individual contributions will be discussed in the following sections, though it is not clear whether these mechanisms work synergistically or in a simple additive fashion.

Regardless of mechanism, toughening is determined by the reduced stress intensity at the crack tip. For crack shielding mechanisms (transformation toughening and microcracking), the local stress intensity, K_l, is "shielded" from the remote field, K_∞, by an amount ΔK:

$$\Delta K = K_\infty - K_l. \tag{5. 1}$$

Crack propagation occurs when the fracture toughness of the material at the tip of crack, K_c^m, is reached. The resultant toughening is determined from the measured toughness, K_c^∞, and the toughness of the transformed/microcracked material:

$$\Delta K_c = K_c^\infty - K_c^m. \tag{5.2}$$

Methods of determining ΔK_c are developed herein.

5.2.1 Transformation Toughening

Since the direct observation of the tetragonal monoclinic transformation in PSZ was made by Porter and Heuer,[11] a number of models have been proposed to explain the stress-induced toughening.[12-14] At present, two equivalent explanations of transformation toughening merit review. The first, by Budiansky et al.,[15] uses a continuum mechanics energy balance; the second, by Marshall et al.,[16] employs the traditional thermodynamic energy balance.

In the former, the energy balance symbolizes the conservation of energy that is applied remotely (the global stress intensity, K_∞) to that used in the near tip region (K_t). For those materials which contain a fully developed transformation zone of height, h, any volume element in the wake has now experienced loading and unloading during steady state crack growth (Fig. 5.1) and must be accounted for in the energy balance shown:

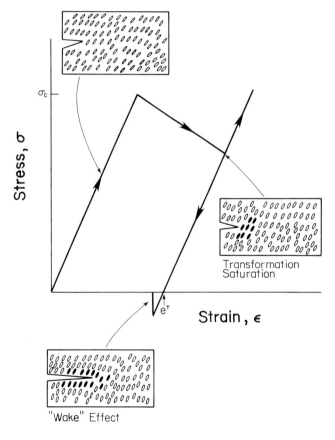

Fig. 5.1 Schematic stress-strain curve of a volume element near the crack tip subject to transformation toughening. The element undergoes loading in the near tip region, transformation at $\sigma = \sigma_c$ and unloading behind the crack tip.

$$K_\infty^2 = K_l^2 + [2E/(1-\nu^2)] \int_0^h U(y)\, dy \tag{5.3}$$

where E is the elastic modulus, ν is Poisson's ratio and $U(y)$ is the residual energy density in the wake. $U(y)$ requires knowledge of the stress-strain response of a volume element which undergoes loading ahead of the crack, transformation at a critical stress and unloading in the wake. Precisely, $U(y)$ is computed from the area enclosed by the stress-strain hysteresis shown in Fig. 5.1:

$$U(y) = \sigma_c e^T V_f + \frac{\bar{B}(e^T V_f)^2}{2(1-\bar{B}/B)} + \frac{E(e^T V_f)^2}{9(1-\nu)} \tag{5.4}$$

where σ_c decribes the mean hydrostatic stress which induces the transformation, e^T is the transformation strain, V_f is the fraction of transforming particles and B, the bulk modulus. The latter term accounts for compressive tractions form the matrix surrounding the wake. $\bar{B}$ describes the stress-strain gradient during transformation.*

If only the frontal zone is considered, the residual energy density term is zero and no toughening results, as K_∞ and K_l are equivalent. By virtue of this argument, the toughening derives strictly from a "wake" effect. Upon evaluation of the residual energy density term in Eq. (5.2) and by substitution of the relationship between zone height and the transformation related stresses:[17]

$$h = [\sqrt{3(1+\nu)^2}/12\pi](K^\infty/\sigma_c)^2, \tag{5.5}$$

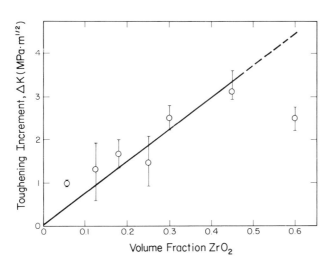

Fig. 5.2 Toughness increases (ΔK_c) for Al_2O_3-ZrO_2 composites. The toughening linearly increases with tetragonal ZrO_2 content. [after F. F. Lange[47]]

the resultant toughening increment can be written as:

$$\Delta K_c = (0.22/1-\nu) E V_f e^T \sqrt{h}. \tag{5.6}$$

The toughening may now be related directly to the amount of transformable ZrO_2

---

* If all transformable ZrO_2 particles undergo transformation, $\bar{B} = -4G/3$, where G is the shear modulus. For $\bar{B} > -4G/3$, $U(y)$, and consequently, the toughening are reduced.

present (Fig. 5. 2), to the zone size which depends on microstructural design,[2)] and to the host material through the elastic modulus term.

Marshall et al.'s[16)] thermodynamic approach similarly sums the energy terms over the complete transformation zone: frontal process zone and wake. The total energy change, dU, of the system which undergoes transformation toughening during the incremental extension of a crack, dc, may be written as:

$$dU = -qdc + 2U_s + d\Phi + dU_D \qquad (5.7)$$

where qdc is the mechanical energy released, $2U_s dc$ is the energy required to create new fracture surfaces, $d\Phi$ is the change in potential energy associated with all of the transformed particles, and dU_D is the energy dissipated when the particles transform. The first two terms of the equation are related to crack propagation in the absence of any crack tip processes. At the critical condition when $q = q_c$, $dU/dc = 0$ and the toughening, Δq_c, resides in the analysis of $d\Phi/dc$ and dU_D/dc. On transformation, the total change in potential may be described by:

$$\Delta\Phi = -\Delta F_{chem} + \Delta U_T - \Delta U_{int} \qquad (5.8)$$

where ΔF_{chem} is the chemical free energy (the driving force for the transformation), ΔU_T is the sum of the strain energy and interface energy, and ΔU_I is the interaction energy (=

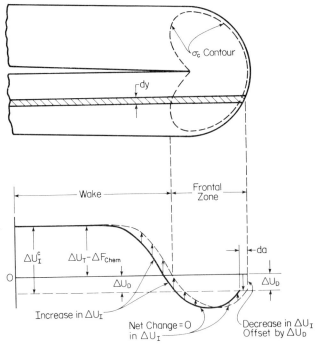

Fig. 5. 3 The change in potential energy with crack advance, dc, for a strip of material, dy, through the frontal zone and wake of a transformation toughened material shown in upper schematic. The change in potential with crack advance is designated by arrows in the lower schematic. The net change in potential is positive from the "wake" contribution. [after D. B. Marshall et al.[16)]]

σe^T). Transformation *ahead* of the crack requires thermodynamic equilibrium (i. e. $\Delta \Phi + \Delta U_D = 0$). Net energy changes derive from the change in potential, primarily the interaction energy, which occurs behind the newly transformed region.

Figure 5. 3 demonstrates the change in the relative potential energy contributions for a strip of material of width, dy, through the frontal zone and wake. From the plot, it is clear to see that the total change in energy is positive, but occurs from the increase in energy *behind* the crack tip. Δg_c is related directly to the interaction energy integrated over the zone width:

$$\Delta g_c = V_f \int_{-h}^{h} U_I dy = 2 V_f e^T \sigma_c h. \tag{5.9}$$

On substitution of Eq. (5. 5), the increase in toughening is equivalent to that predicted in Eq. (5. 6).

By comparing measured toughening data with Eq. (5. 6), one finds that the toughening increment is consistently underestimated. For example, Eq. (5. 6) predicts a ΔK_c of 0.81 MPa m$^{1/2}$ for PSZ studied by Porter and Heuer. Observed increases in toughness are on the order of 2.3 MPa m$^{1/2}$. Clearly, the Budiansky et al. and Marshall et al. approaches cannot account for the large magnitude of the toughening increases observed, and additional transformation-related effects are under consideration.

5.2.2 Microcracking

Transmission microscopy of transformation toughened materials suggests that microcracking accompanies the stress-induced martensitic transformation in numerous systems.[18,19] In addition, stress-induced microcracking may also occur in materials containing monoclinic zirconia which has transformed during processing.[20-22]

A recent analysis[23] describes the toughening effect of both microcracking situations using a similar continuum energy balance to that proposed by Budiansky et al.[15] (cf. Subsec. 5. 2. 1.) Consider the changes in the stress-strain hysteresis when microcracking accompanies the transformation. (Fig. 5. 4) On microcrack formation, an increase in strain is observed. Simultaneously, a change in compliance results, evidenced by the a non-parallel unloading path behind the crack tip. The latter modulus reduction may counter-balance the strain increase and *may* give rise to a permanent strain, θ, less than e^T. Thus, a toughness enhancement or reduction is possible. The work density term, $U(y)$ is modified as follows:

$$U(y) = \sigma_c \theta f_s + \frac{\sigma_c^2}{2}\left(\frac{1}{B_1} - \frac{1}{B}\right) + \frac{\overline{B}\left[\theta f_s \sigma_c \left(\frac{1}{B_1} - \frac{1}{B}\right)\right]^2}{2\left[1 - \overline{B}/B_1\right]}$$

$$+ \frac{B_1(1 - 2\nu_1)\theta^2 f_s^2}{3(1 - \nu_1)} \tag{5.10}$$

where the second and third terms now account for the modulus reduction. The subscript, 1, refers to the microcracked material and f_s to the volume fraction of microcracks. Upon integration of Eq. (5. 10) and substitution of Eq. (5. 5), an analytic expression for

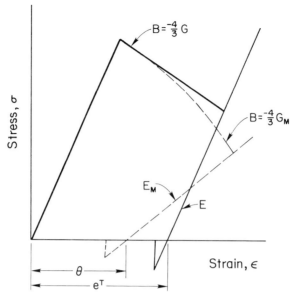

Fig. 5. 4 Idealized stress-strain hysteresis for volume element undergoing microcracking along with the stress-induced transformation. [after K. T. Faber[23]]

microcracking results:

$$\Delta K \cong 0.25 E f_s \theta \sqrt{h}. \tag{5.11}$$

The net effect of microcracking (with the transformation) on toughening requires a knowledge of the microcrack density and microcrack opening displacement which allow computation of both strain and compliance changes. Figure 5. 5 delineates regions of

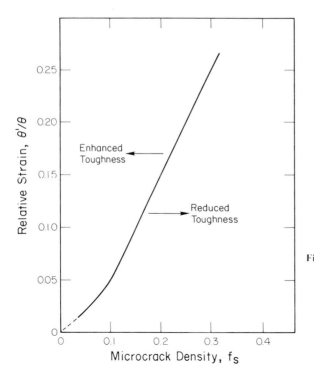

Fig. 5. 5 Normalized increased strain vs. microcrack density to define conditions for enhanced or reduced toughening when microcracking accompanies the martensitic transformation. [after K. T. Faber[23]]

toughness enhancement or reduction for relative values of strain increase to modulus reduction.

Eq. (5. 11) also holds for microcracking in the absence of the stress-induced transformation.[24] MgO stabilized ZrO_2 containing fine monoclinic particles produced from a sub-eutectoid heat treatment demonstrate $\Delta K_c = 6$.[25] Similar observations have been made in the Al_2O_3-ZrO_2 systems.[22]

Note that transformation toughening and microcrack toughening theoretically provide equivalent toughening by comparing Eqs. (5. 6) and (5. 11). It is unlikely, however, that the transformation and microcracking strains would be identical. In general, $\theta < e^T$.

5.2.3 Crack Deflection

In addition to the crack shielding methods that have been proposed above, the transformed ZrO_2 particles and their associated volume expansion act as a source of residual stresses. The local residual stresses may determine the path of crack propagation and give rise to crack deflection toughening. Regions where a crack is not planar (and not normal to the direction of applied stress) comprise crack deflections. Such increments diminish the stress intensity, and consequently, serve to toughen the material.

Locally, the crack driving force is determined by the local stress intensities at each deflected segment of the crack front. Advance of each increment is then governed by the local strain energy release rate, q:

$$E q = k_1^2(1-\nu^2) + k_2^2(1-\nu^2) + k_3^2(1+\nu) \tag{5.12}$$

where the lower case k refers to the local stress intensity. The average strain energy release rate, $<q>$, across the crack front may then represent the crack driving force. The toughening may be calculated by comparing $<q>$ with the strain energy release rate for an undeflected crack, q_m:

$$q_c = (q^m / <q>) q_c^m \tag{5.13}$$

To assess the average strain energy release rate, a two step process is used. First, the stress intensities for non-planar cracks are formulated and depend on the character of the deflection (i. e. Mode II and Mode III components). Second, the distribution of deflection angles are determined by geometric and statistical arguments for the specific particle size distributions and morphologies. As demonstrated by the Faber and Evans analysis,[26] the magnitude of the toughening increase is dependent on the volume fraction of the deflecting phase, and more importantly, the morphology. Elongated particles, such as discs and rods, provide greater toughening. The magnitude of ΔK_c increases with aspect ratio.(Fig. 5. 6)

From the crack deflection model, the toughening which arises from ZrO_2 particles having an aspect ratio (L/d) of approximately 4, present in volume fractions of approximately 30% should result in $K_c/K_c^m = 1.5$ (Fig. 5. 6) Toughness measurements of monoclinic ZrO_2 in ZnO yield similar results.[27] In Figure 5. 7, overaged ZrO_2-particles in the CaO-stabilized system serve as crack deflectors, where the aspect ratio increases during aging. L/d may now approach 20 and $K_c/K_c^m = 2$. Comparative toughness measurements are not available.

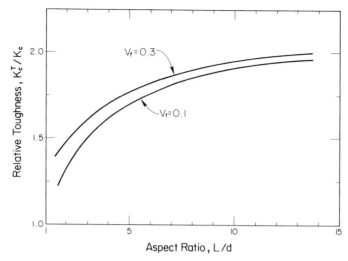

Fig. 5. 6 Toughening predictions as a function of aspect ratio for crack deflection around rod-shaped particles contained in volume fractions of 0.10 and 0.30.

Fig. 5. 7 Scanning electron micrograph of crack deflection around coarse precipitates in overaged MgO partially stabilized ZrO_2. [after M. V. Swain and R. J. H. Hannink[4]]

Although the crack shielding phenomena described earlier is generally thought to provide substantially greater toughening than crack deflection processes, crack deflection toughening should be operative at much higher temperatures than either transformation or microcrack toughening. This method would be responsible for maintaining enhanced fracture toughness at elevated temperatures, since residual stress fields would still exist in a transformed or untransformed system.

5.2.4 R-Curve Behavior

Since toughening in crack shielding systems derives primarily from wake effects, it follows that the crack growth resistance should be depend upon the extent of crack advance (R-curve behavior). Maximum toughening should be afforded only when a fully developed wake is present. (Fig. 5.8) Our physical understanding of R-curve behavior supercedes any theoretical formulation for the shape of the R-curve. One proposed analytic fit for the R-curve is[24]

$$K_R/K_c \propto \tan^{-1}(\Delta a/h). \qquad (5.14)$$

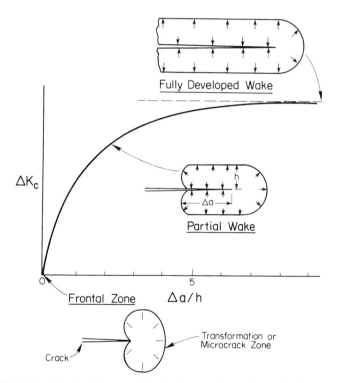

Fig. 5.8 Predicted R-curve behavior for transformation toughening or microcracking, demonstrating asymtotic behavior at $\Delta a \simeq 5h$.

A plot of the above equation suggests that most of the toughening is seen when $\Delta a \cong 5h$. Measured R-curves for some ZrO_2-based materials generally support this idea. R-curve measurements by Swain et al.[4] suggest a zone height of approximately $5\mu m$, well within the error of the direct observations of the transformation zones.(Fig. 5.9)

R-curve effects are also evidenced when comparing results from a variety of fracture

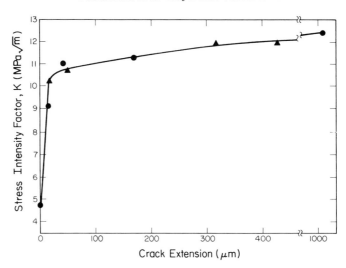

Fig. 5. 9 R-curve from notched-beam test of partially stabilized ZrO$_2$. [after M. V. Swain and R. J. H. Hannink[4)]]

toughness techniques. In Fig. 5. 10 are plotted a series of fracture toughness data for Al$_2$O$_3$-ZrO$_2$, annealed and unannealed, and tested using two fracture toughness test techniques.[28)] For test techniques in which no stable crack growth occurs before fracture, the toughness of the material is determined by K_R at the start of the R-curve. However, if the wake is fully developed before failure, through a machining process or stable crack growth, the asymtotic

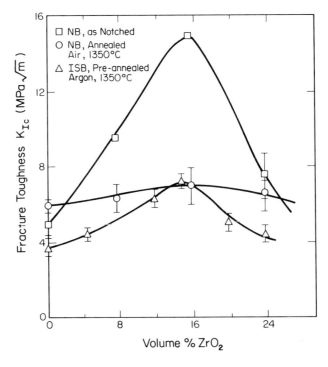

Fig. 5. 10 Fracture toughness of Al$_2$O$_3$-ZrO$_2$ composites with increasing volume fraction of ZrO$_2$ for two test techniques and annealing conditions. [after M. V. Swain and N. Claussen[28)]]

or maximum value in toughness is observed. The consequences of R-curve effects may be manifested during thermal cycling or post-machining annealing under conditions where the monoclinic → tetragonal transformation would occur. Little research has been directed toward this problem of great significance.

5.3 Transformation-Toughened Surfaces

Though fracture toughness is a property associated with a bulk material response, the transformation toughening phenomenon can be used to achieve compressive surface layers which impart an "effective" toughening upon the material. The phenomenon is related to one first observed in glass, where thermal or chemical tempering results in strength increases on the order of 50 to 100%.[29] In the ZrO_2-based materials, the compressive layer is achieved by transforming only the near-surface tetragonal particles to their monoclinic form. Because of the volume expansion that accompanies the crystallographic transformation, the surface layer possesses a larger effective volume than the bulk, just as in the case of tempered glass, giving rise to a surface compressive layer.

The surface transformation was originally (and most commonly) observed after machining, where strength increases as great as 100% were observed.[30-33] Most recently, however, Green et al.[34,35] have utilized a thermochemical treatment to create the surface tetragonal → monoclinic transformation. By preferential removal of Y_2O_3 or CeO_2 stabilizers from the surface layer through a high temperature treatment, transformation takes place on cooling but only in Y_2O_3- or CeO_2-depleted regions.

The extent of the "effective" toughening by surface treatments can be seen from Green's measure of the radial cracks from Vickers hardness impressions used for toughness measurements since[34,36]

$$K_c \propto c^{-3/2} \tag{5.15}$$

and c is the crack length. (Table 5.2) Residual compressive stresses on the order of 550 MPa may result.

Table 5.2 Comparison of indentation crack sizes residual stresses for as-fabricated and heat-treated surfaces of Y_2O_3-ZrO_2. [after D. J. Green[34]]

Surface Treatment	Radial Crack* Radius (μm)*	Residual Stress (MPa)
As Fabricated	134	
Y_2O_3 Removal 1,400°C/1 hour	91.6	−140
1,400°C/4 hours	82.3	−330
1,400°C/16 hours	80.0	−550

*Indentation made with 98 N load.

By the thermochemical treatment, the temperature sensitivity (Subsec. 5.4.2) of the surface "toughening" is diminished. The phenomenon should retain its effectiveness to temperatures near the transformation range. This development implies that these materials will be useful at high temperatures and represents a significant contribution by Green and his colleagues. However, the "effective" toughening resulting from surface treatments will *only* be useful if a material is subjected to large surface (contact) stresses and fails from surface or sub-surface related flaws located within the transformed (compressed) layer.

5.4 Environmental Effects

The toughening models derived in Section 5.2, though providing a physical interpretation of how toughening is achieved, provide no insight into temperature or environmental effects. As an example, although a size dependence of retaining tetragonal ZrO_2 to room temperature is recognized, knowledge of ZrO_2 size and size distribution does not necessarily allow computation of toughening degradation with grain coarsening at elevated temperature. Critical sizes for spontaneous transformation depend upon the stabilizer, host material, processing method and precipitate morphology.[2] At present, our knowledge and potential for modelling environmental effects is limited by the sparcity of experimental observations. They are reviewed here.

5.4.1 Temperature Effects

The chemical driving force, ΔF_{chem}, is the only temperature dependent function affecting toughening. It is well established that this driving force decreases with temperature. In recent work by Lange, a linear dependence between the driving force and toughness has been established.[37] In Fig. 5.11, data collected by Lange on Al_2O_3-ZrO_2's demonstrates the predicted decline in the fracture toughness up to temperatures of 600°C.* Note, however, that at elevated temperatures, the fracture toughness of the Al_2O_3 host is still surpassed, despite a decrease in toughness by more than 30%. At elevated temperatures the toughening must derive from less temperature sensitive machanisms such as crack deflection.

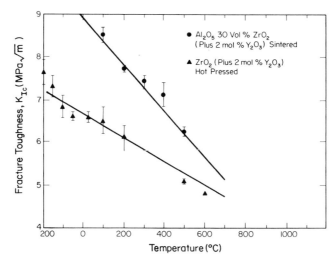

Fig. 5.11 Fracture toughness plotted against temperature for ZrO_2 and Al_2O_3-ZrO_2. [after F. F. Lange[37]]

---

* K_{IC} continues to increase linearly below room temperature to $-200°C$.[37,38]

5.4.2 Aging Effects

Aging studies have been performed above and below the transformation temperature where significant changes in mechanical behavior are widely observed. In the first, above the eutectoid temperature, grain coarsening may occur, in which the stabilizer is expelled through a decomposition reaction. Tetragonal particles are converted to the monoclinic form on cooling. Studies by Bhathena et al.[39] clearly demonstrate these aging effects in a unique manner. In a MgO partially stabilized ZrO_2 double cantilever beam specimens were heat treated in a temperature gradient. Crack growth behavior was monitored over an extensive aging temperature profile in a single sample. (Fig. 5. 12) Both time and temperature effects to provide optimally aged microstructures are discernable from this data.

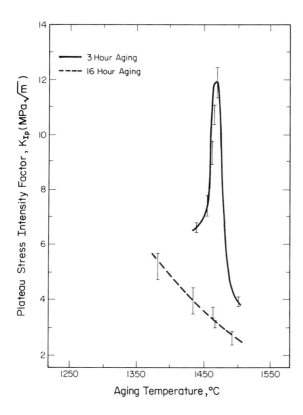

Fig. 5. 12 Plateau stress intensity plotted against annealing temperature for a MgO partially stabilized ZrO_2 for two annealing treatments. [after N. Bhathena et al.[39]]

Sub-eutectoid aging in the MgO-containing ZrO_2 system has been well established by Swain et al.[40] where cubic stabilized ZrO_2 directly decomposes to monoclinic ZrO_2 and MgO. The decomposition at short times (< 5 hours) is accompanied by extensive mocrocracking of the cubic matrix, where linking microcracks degrade mechanical properties. At longer decomposition times, only monoclinic ZrO_2 remains along with a network of MgO "pipes". Enhanced toughness is due to stress-induced microcracking.

A second sub-eutectoid treatment, of more recent interest, has only been observed in the Y_2O_3-ZrO_2 system. The degradation appears in the temperature regime of 100 to 400°C at annealing times as short as 5 hours, whereupon the surface of the annealed materials is

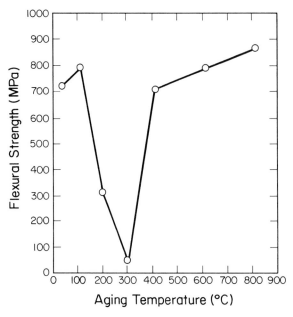

Fig. 5. 13 Flexural strength for tetragonal zirconia polycrystals containing 4 mol% Y_2O_3 after 100 h treatment. [after M. Watanabe et al.[41]]

characterized by severe microcracking and the materials suffer strength degradation.[41-43] (Fig. 5. 13)

Originally, the phenomenon was thought to be a stress corrosion cracking phenomenon in materials containing an amorphous grain boundary phase, characteristic of the Y_2O_3-containing materials. Recent evidence by Lange et al.[44] indicates that the phenomenon in entirely moisture-related and corresponds to the preferential reaction of the yttrium with water, regardless of the amount of glass present at the boundaries. The problem is circumvented by use of extremely fine-grained tetragonal ZrO_2 precipitates which are not subject to transformation to the monoclinic form.[45] (Fig. 5. 14.)

5.4.3 Subcritical Crack Growth

Studies of slow crack growth in ZrO_2-based materials have been limited. Becher[46] has characterized the slow crack behavior of Al_2O_3-ZrO_2's containing Y_2O_3 and found an improved slow crack growth resistance compared to non-transformation-toughened alumina composites. (Fig. 5. 15) The stress corrosion exponent, n, of the Al_2O_3-ZrO_2 materials is significantly lower than that for fine-grained Al_2O_3. It is speculated that residual stresses associated with large transformed particles are the cause. No correlation was made between the slow crack growth behavior and the presence of an amorphous phase ubiquitous to conventionally-processed Y_2O_3-ZrO_2's.

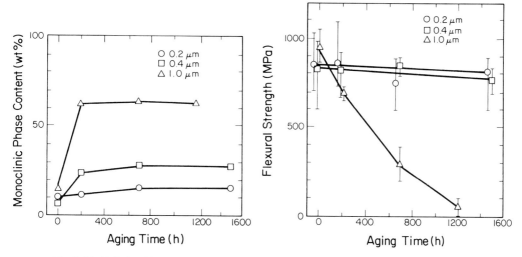

Fig. 5.14 Relationship between grain size and 230°C aging time for tetragonal zirconia polycrystals containing 3 mol% Y_2O_3. Left: grain size dependence of phase transformation on the surface. Right: grain size dependence of fracture strength. [after K. Tsukuma et al.[45]]

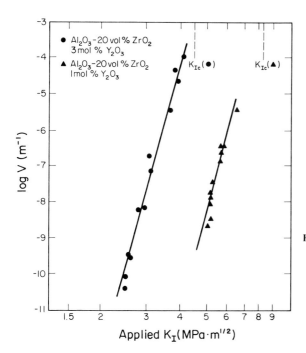

Fig. 5.15 Slow crack growth behavior of Al_2O_3-20 vol% ($ZrO_2 \cdot xY_2O_3$), demonstrating the enhanced slow crack growth resistance with transformation toughening. [after P. F. Becher[46]]

5.5 Summary and Conclusions

A review concentrated on the mechanics of toughening and the mechanical properties of ZrO_2-based materials has been provided. Three toughening mechanisms have been discussed, transformation toughening, microcracking and crack deflection, which form a framework to describe how crack growth resistance is enhanced. Though the models cannot explain the absolute magnitude of toughness enhancement, they do provide *some* guidelines for microstructural design. Clouding our ability to predict transformation-toughening behavior, however, are the variety temperature and environmental effects which enhance or degrade mechanical behavior: temperature, aging effects, humidity and slow crack growth, to name a few. The paucity of data makes this field ripe for study to improve our understanding of the many aspects of transformation toughening.

Acknowledgements

The author acknowledges support by the NSF Presidential Young Investigator Award (DMR-8351476) in conjunction with IBM Corporation, 3M Corporation, Eastman Kodak and PPG Industries.

References

1) R. C. Garvie, R. H. Hannink and R. T. Pascoe: "Ceramic steel?," Nature (London), 258, 5537 (1975) 703-704.
2) N. Claussen: "Microstructural design of zirconia-toughened ceramics (ZTC)," in Advances in Ceramics, Vol. 12, ed. by N. Claussen, M. Ruhle and A. H. Heuer, American Ceramic Society, Columbus, Ohio (1984) 325-351.
3) W. M. Kriven: "Shear transformations in inorganic materials," in Solid-Solid Phase Transformations, ed. by H. I. Aaronson, D. E. Laughlin, R. F. Sekerka and C. M. Wyman, The Metallurgical Society of AIME, Warrendale, PA (1982) 1507-1531.
4) M. V. Swain and R. J. H. Hannink: "R-curve behaviour in zirconia ceramics", in Advances in Ceramics, Vol. 12, ed. by N. Claussen, M. Ruhle and A. H. Heuer, American Ceramic Society, Columbus, Ohio (1984) 225-239.
5) P. F. Becher and V. J. Tennery: "Fracture toughness of Al_2O_3-ZrO_2 composites," in Fracture Mechanics of Ceramics, Vol. 6, ed. by R. C. Bradt, A. G. Evans, D. P. H. Hasselman and F. F. Lange, Plenum Publishing Corp., New York (1983)383-399.
6) F. F. Lange, B. I. Davis and D. O. Raleigh: "Transformation strengthening of beta″-Al_2O_3 with tetragonal ZrO_2," J. Am. Ceram. Soc., 66, 3 (1983) C50-52.
7) L. J. Gauckler, J. Lorenz, J. Weiss and G. Petzow: "Improved fracture toughness of SiC-based ceramics," in Science of Ceramics, Vol. 10, ed. by H. Hausner, Academic Press, London (1980) 577-584.
8) F.F.Lange: "Compressive surface stresses developed by ceramics by an oxidation-induced phase change," J. Am. Ceram. Soc., 63, 1 (1980) 38-40.
9) M. A. Choudhry and A. G. Crocker: "Theory of twinning and transformation modes in zirconia," in Advances in Ceramics, Vol. 12, ed. by N. Claussen, M. Ruhle and A. H. Heuer, American Ceramic Society, Columbus, Ohio (1984) 46-53.
10) M. Ruhle and A. H. Heuer: "Phase transformations in ZrO_2-containing ceramics: II," in Advances in Ceramics, Vol. 12, ed. by N. Claussen, M. Ruhle and A. H. Heuer, American Ceramic Society, Columbus, Ohio (1984) 14-32.
11) D. L. Porter and A. H. Heuer: "Mechanisms

of toughening partially stabilized zirconia (PSZ)," J. Am. Ceram. Soc., **60**, 5-6 (1977) 183-184.
12) A. G. Evans and A. H. Heuer: "REVIEW—Transformation toughening in ceramics," J. Am. Ceram. Soc., **63**, 5 (1980) 241-248.
13) F. F. Lange: "Transformation toughening, Part 2," J. Mater. Sci., **17**, 1 (1982) 235-239.
14) S. J. Burns and J. R. Michener: "Thermodynamics of phase transformation toughening," in Fracture Mechanics of Ceramics, Vol. 6, ed. by R. C. Bradt, A. G. Evans, D. P. H. Hasselman and F. F. Lange, Plenum Publishing Corp., New York (1983) 275-287.
15) B. Budiansky, J. W. Hutchinson and J. C. Lambropoulos: "Continuum theory of dilatant transformation toughening in ceramics," Int. J. Solids Struct., **19**, 4 (1983) 337-355.
16) D. B. Marshall, A. G. Evans and M. Drory: "Transformation toughening in ceramics," in Fracture Mechanics of Ceramics, Vol. 6, ed. by R. C. Bradt, A. G. Evans, D. P. H. Hasselman and F. F. Lange, Plenum Publishing Corp., New York (1983) 289-307.
17) R. M. McMeeking and A. G. Evans: "Mechanics of transformation toughening in brittle materials," J. Am. Ceram. Soc., **65**, 5 (1982) 242-246.
18) Y. Fu, A. G. Evans and W. M. Kriven: "Microcrack nucleation in ceramics subject to a phase transformation," J. Am. Ceram. Soc., **67**, 9 (1984) 626-630.
19) M. Ruhle, A. Strecker, D. Waidelich and B. Kraus: "In situ observations of stress-induced phase transformations in ZrO_2-containing ceramics," in Advances in Ceramics, Vol. 12, ed. by N. Claussen, M. Ruhle and A. H. Heuer, American Ceramic Society, Columbus, Ohio (1984) 256-274.
20) N. Claussen: "Fracture toughness of Al_2O_3 with an unstabilized ZrO_2 dispersed phase," J. Am. Ceram. Soc., **59**, 1-2 (1976) 49-51.
21) D. J. Green, P. S. Nicholson and J. D. Embury: "Fracture toughness of partially stabilized ZrO_2 in the system $CaO-ZrO_2$," J. Am. Ceram. Soc., **56**, 12 (1973) 619-623.
22) N. Claussen, R. L. Cox and J. S. Wallace: "Slow growth of microcracks: Evidence for one type of ZrO_2 toughening," J. Am. Ceram. Soc., **65**, 11 (1982) C190-191.
23) K. T. Faber: "Microcracking contributions to the toughness of zirconia-based ceramics," in Advances in Ceramics, Vol. 12, ed. by N. Claussen, M. Ruhle and A. H. Heuer, American Ceramic Society, Columbus, Ohio (1984) 293-305.
24) A. G. Evans and K. T. Faber: "Crack-growth resistance of microcracking brittle materials," J. Am. Ceram. Soc., **67**, 4 (1984) 255-260.
25) M. V. Swain: "R-curve behaviour of magnesia partially stibilized zirconia and its significance to thermal shock," in Fracture Mechanics of Ceramics, Vol. 6, ed. by R. C. Bradt, A. G. Evans, D. P. H. Hasselman and F. F. Lange, Plenum Publishing Corp., New York (1983) 355-369.
26) K. T. Faber and A. G. Evans: "Crack deflection processes—I. Theory," Acta Metall., **31**, 4 (1983) 565-576.
27) H. Ruf and A. G. Evans: "Toughening by monoclinic zirconia," J. Am. Ceram. Soc., **66**, 5 (1983) 328-332.
28) M. V. Swain and N. Claussen: "Comparison of K_{IC} values for Al_2O_3-ZrO_2 composites obtained from notched-beam and indentation strength techniques," J. Am. Ceram. Soc., **66**, 2 (1983) C27-29.
29) D. G. Holloway: The Physical Properties of Glass, Wykeham Publications Ltd., London (1973) 190-196.
30) R. T. Pascoe and R. C. Garvie: "Surface strengthening of transformation toughened zirconia," in Ceramic Microstructures '76, ed. by R. M. Fulrath and J. A. Pask, Westview Press, Boulder, CO (1977) 774-784.
31) J. S. Reed and A. Lejus: "Effect of granding and polishing on near-surface phase transformations in zirconia," Mater. Res. Bull., **12** 10 (1977) 949-954.
32) T. K. Gupta: "Strengthening by surface damage in metastable tetragonal zirconia," J. Am. Ceram. Soc., **63**, 1-2 (1980) 117.
33) N. Claussen and G. Petzow: "Strengthening and toughening models in ceramics based on ZrO_2 inclusions," in Energy and Ceramics, ed. by P. Vincenzini, Elsevier, Amsterdam (1980) 680-691.
34) D. J. Green: "A technique for introducing surface compression into zirconia ceramics," J.

Am. Ceram. Soc., **66**, 10 (1983) C178-179.

35) D. J. Green, F. F. Lange and M. R. James: "Residual surface stresses in Al_2O_3-ZrO_2 composites," in Advances in Ceramics, Vol.12, ed. by N. Claussen, M. Ruhle and A. H. Heuer, American Ceramic Society, Columbus, Ohio (1984) 240-250.

36) D. B. Marshall and B. R. Lawn: "An indentation method for measuring stress in tempered glass," J. Am. Ceram. Soc., **60**, 1 (1977) 86-87.

37) F. F. Lange: "Transformation toughening, Part 5. Effect of temperature and alloy on the fracture toughness," J. Mater. Sci., **17**, 1 (1982) 255-262.

38) C. A. Anderson and T. K. Gupta: "Phase stability and transformation toughening in zirconia," in Advances in Ceramics, Vol.3, ed. by A. H. Heuer and L. W. Hobbs, American Ceramic Society, Columbus, Ohio (1981) 184-210.

39) N. Bhathena, R. G. Hoagland and G. Meyrick: "Effects of particle distribution on transformation-induced toughening in MgO-PSZ," J. Am. Ceram. Soc., **67**, 12 (1984) 799-805.

40) M. V. Swain, R. C. Garvie, R. H. J. Hannink: "Influence of thermal decomposition on the mechanical properties of magnesia stabilized cubic zirconia," J. Am. Ceram. Soc., **66**, 5 (1983) 358-562.

41) M. Watanabe, S. Iio and I. Fukuura: "Aging behavior of Y-TZP," in Advances in Ceramics, Vol.12, ed. by N. Claussen, M. Ruhle and A. H. Heuer, American Ceramic Soiety, Columbus, Ohio (1984) 391-398.

42) T. Sato and M. Shimada: "Crystalline phase-change in yttria-partially-stabilized zirconia by low-temperature annealing," J. Am. Ceram. Soc., **67**, 10 (1984) C212-213.

43) K. Kobayashi, H. Kuwajima and T. Masaki: "Phase change and mechanical properties of ZrO_2-Y_2O_3 solid electrolyte after aging," Solid State Ion., **3-4** (1981) 489-493.

44) F. F. Lange, G. Dunlop and B. I. Davis: to be published.

45) K. Tsukuma, Y. Kubota and T. Tsukidate: "Thermal and mechanical properties of Y_2O_3-stabilized tetragonal zirconia polycrystals," in Advances in Ceramics, Vol.12, ed. by N. Claussen, M. Ruhle and A. H. Heuer, American Ceramic Society, Columbus, Ohio (1984) 382-390.

46) P. F. Becher: "Slow crack-growth behavior in transformation-toughened Al_2O_3-ZrO_2 (Y_2O_3) ceramics," J. Am. Ceram. Soc., **66**, 7 (1983) 485-448.

47) F. F. Lange: "Transformation toughening, Part 4. Fabrication, fracture-toughness and strength of Al_2O_3-ZrO_2 composites," J. Mater. Sci., **17**, 1 (1982) 247-254.

6. A Review of Creep in Silicon Nitride and Silicon Carbide

Robert F. DAVIS* and Calvin H. CARTER, Jr.*

Abstract

The processing routes used to fabricate Si_3N_4 and SiC are similar, e.g., reaction-bonding, sintering and hot pressing with the last two procedures requiring additives to achieve densification. However, the crystal structures and the microstructures resulting from the aforementioned processes are markedly different. As such, creep in reaction-bonded Si_3N_4 is controlled primarily by processes which depend on the presence of oxide formation; a similarly prepared SiC creeps by climb-controlled dislocation glide and climb processes. The creep mechanism in hot-pressed Si_3N_4 is dependent on the amount of noncrystalline grain boundary phase: substantial amounts lead to cavitation as the principal mechanism, while materials with small amounts deform by diffusion and reprecipitation processes occurring in the boundary phase. In contrast, initial results from creep studies in sintered SiC having B- or Al-containing additives indicate that lattice or grain boundary diffusion mechanisms are the controlling processes. The research which has led to these findings and conclusions is described in this manuscript.

Keywords: silicon nitride, silicon carbide, creep, electron microscopy, deformation mechanisms.

6.1 Introduction

Silicon nitride (Si_3N_4) and silicon carbide (SiC) have recently been at the forefront of developments in high-strength, high temperature materials, as they have been the focus of an international effort to develop more efficient, ceramic-based gas turbine and diesel engines which would be operative at temperatures above the useful range for metal superalloys. In addition, one or both of these materials is being used or proposed for employment in a host of other energy conversion or energy transfer devices which would be employed at high temperatures. As such, deformation of these materials as a function of stress, temperature and time is of concern for the prolonged utilization of these devices. A substantial number of deformation/creep studies have been conducted on Si_3N_4 materials prepared by several process routes; however, relatively few studies of this type have been carried out on the similarly diverse forms of SiC. The following sections present a digest of

---

* Department of Materials Engineering, Box 7907, North Carolina State University, Raleigh, North Carolina 27695-7907, U. S. A.

the most recent deformation studies. However, reference to some of the earlier research is also given. Finally, this review is particularly concerned with the kinetics and mechanisms of creep in the steady-state regime. Studies of transient creep or lifetime prediction are not described.

6.2 Silicon Nitride

6.2.1 Structure and Polymorphism

As in silicates, the silicon in Si_3N_4 is considered as an excited state sp^3 configuration. Such a configuration produces the usual tetrahedral arrangement of valence orbitals for covalent bond formation with 4 nitrogen atoms. These SiN_4 tetrahedra form a three-dimensional network by sharing corners such that each N is common to three tetrahedra (in contrast to two tetrahedra for O in SiO_2). The greater rigidity of the tetrahedral bridging units in Si_3N_4 gives rise to the outstanding properties of the material and the dominance of one crystal structure, namely the hexagonal β-Si_3N_4 form. This dominant form of Si_3N_4 has the designation "β" because for some time an alternative form designated α has been described and its composition originally disputed. This dispute centered around whether α is a form of Si_3N_4 or whether it is a ternary compound containing oxygen atoms and cation vacancies.[1] However considerable evidence[2-6] now shows that the α-structure to be pure Si_3N_4 and that the oxygen is present as either SiO_2, Si_2N_2O, cation impurity oxides or complex silicates.

6.2.2 Processing Routes

Silicon nitride is a generic name for a material which is fabricated by several process routes which result in a variety of microstructures and associated properties. Although Si_3N_4 powders are relatively easy to obtain, the material contains few intrinsic vacancies even at high temperatures and is not given to plastic flow. Therefore the common densification mechanisms of solid state sintering are not readily available, and the material cannot be sintered to high density by the application of heat alone. The most common powder processing routes include hot pressing, which requires oxide additives to achieve near theoretical densities and reaction bonding. These procedures will be briefly described in the following paragraphs.

In hot pressing, the powdered α-Si_3N_4 is mixed with densification aids such as MgO, MnO, and Y_2O_3 and heated to 1,900–2,150 K at 20 MN/m² pressure in a graphite die. After approximately thirty minutes, β-Si_3N_4 artefacts of very high density are produced. The function of the additives is to react with the outer layer of SiO_2 on each particle to provide a liquid phase of sufficiently low viscosity to enable densification before the SiO_2 can react with the Si_3N_4 to form a crystalline oxynitride. Densification occurs by three stages: (1) particle rearrangement; (2) solution of α-Si_3N_4 and reprecipitation of β-Si_3N_4; and (3) coalescence.

MgO was the first widely used additive. Reaction with the surface SiO_2 produces a silicate liquid near to the $MgSiO_3$–SiO_2 eutectic composition if small amounts ($\sim$ 2%) of MgO are used. The eutectic liquid reacts with Si_3N_4 leading to Mg–Si–N–O glass which easily wets and dissolves more α-Si_3N_4. Precipitation of β-Si_3N_4 leads to densification. Other impurities, in particular alkali and alkaline earth oxides, lower the viscosity still further and so aid the process. However, on completion there is an intergranular glassy phase (see Refs. 17) and 21-24) for in-depth discussions of this phenomena).

Reaction between SiO_2, Y_2O_3 and Si_3N_4 produces a low viscosity, reactive liquid at 1,973 K which enables densification. There is a final strength improvement with this additive caused by the liquid reacting further with Si_3N_4 to produce a crystalline refractory grain boundary phase, instead of a glass, which means that viscous shear in the boundaries when products are stressed is a reduced problem. Unfortunately, if the overall glass composition is outside the compositional triangle bounded by Si_3N_4, Si_2N_2O, and Si_2O_7, the products of the crystallization process may be cristobalite and $Y_2Si_2O_7$. The formation of these phases causes a marked volume increase and resultant extensive cracking in the grain boundary materials which reduces the strength and oxidation resistance of the material.

The use of approximately 5 wt% ZrO_2 allows the formation of a liquid with the SiO_2 layer that dissolves the α-Si_3N_4. The final products are mixtures lying in the compatibility triangle Si_3N_4-Si_2N_2O-ZrO_2 with only small amounts of grain boundary phase and good oxidation resistance.

The reaction bonding process has the advantage that complex shapes can be preformed by compacting Si powder to moderate density via such methods as cold isostatic pressing, flame spraying, slip casting, injection or warm die molding on thin film techniques. The shaped article is then heated in N_2 at temperatures up to 1,750 K for times varying from 2-5 days, depending on the process and the size of the components, to produce a mixture of α- and β-Si_3N_4. There is very little change in overall dimensions of the compact during the reaction in spite of the fact that the volume of solid increases by 22%. This means that as the reaction continues into the silicon particles it must be accompanied by creation of space so that final porosities of 20-25% are common.

Addition of 2% of metal fluoride to the silicon enhances both the growth rate and α/β ratio by a mechanism that involves devitrification of the silica coating.

Although there are additional Si_3N_4 processing routes, the materials investigated for their deformation and creep behavior have been prepared almost exclusively by hot pressing or reaction bonding. As such, the following review is divided according to the deformation properties of materials prepared by these methods. Because of space limitations, this review will not cover deformation studies in the Si_3N_4 solid solutions known collectively as SiAlONS.

6.2.3 Deformation and Creep
(a) Hot-pressed Si_3N_4

As may be surmised from the foregoing processing information, any review of creep in hot pressed Si_3N_4 materials fabricated to date must focus on the physicochemical processes occurring at the oxide additive-Si_3N_4 interface and within the grain boundary phase rather than in the interior of each grain. Thus it is instructive to initially consider creep research in which the investigators have also taken into account the cogent oxide additive-SiO_2-Si_3N_4 phase equilibria and the resulting microstructures. A major portion of this integrated research has been conducted by Lange and coworkers.[7-16] A review of the results of this work as well as the interrelated analytical TEM research is given in the following paragraphs. The results of earlier research are then examined with reference to the applicable phase equilibria.

As implied in Subsection 6.2.2, oxygen is by far the most abundant impurity in "pure" Si_3N_4 powders. It is present as either SiO_2 or Si_2N_2O. Thus the general reactions during sintering may be expressed[14]

$$Si_3N_4 + SiO_2 + \text{additive (metal oxide)} + \text{other impurities} \xrightarrow{heat} Si_3N_4(S) + \text{Liquid} \quad (6.1)$$

$$Si_3N_4(S) + \text{Liquid} \xrightarrow{cooling} \text{Dense } Si_3N_4 + \text{second phase(s)} \quad (6.2)$$

An examination of the cogent phase equilibria will illustrate the polyphase nature of hot pressed Si_3N_4. Figure 6.1 shows two subsolidus phase diagrams containing the densification additives of MgO and Y_2O_3, respectively. For compositions prepared in either of these two systems, the resulting material could either contain two phases, i.e., compositions reaching equilibrium on a tie line, or three phases, i.e., compositions reaching equilibrium in one of the compatibility triangles containing Si_3N_4 as an end member. If impurities are included, the situation becomes even more complicated. When one fabricates material close to the Si_3N_4 corner of any of these systems, small changes in any of the other constituents can dramatically change the secondary phase. For example, small changes in the oxygen content (or SiO_2 content) can shift the end-point composition from one compatibility triangle to another. Thus control of the oxygen content in the starting powder and during fabrication (control of volatilization reactions involving SiO) is crucial for controlling the type and amount of the secondary phase.

Equilibrium may be obtained during the high temperature densification process. However, it is not achieved during cooling, since, the liquid which forms during densification is largely siliceous. As such, solidification invariably results in a residual, continuous glassy phase which is chiefly responsible for high temperature mechanical properties. Despite the nonequilibrium nature of these materials, the equilibrium phase diagram can still be employed to estimate the composition and volume content of the glassy phase. If one assumes[11] that the composition of the glassy phase is similar to the eutectic composition of the compatibility triangle in which the composition was fabricated, Lange[10] has noted that the content of the glass will depend on composition in the manner described by rules of phase equilibria. In addition, the volume fraction of the glassy grain boundary phase will increase as the composition of the alloy is shifted toward the eutectic composition.

Lange and co-workers have fabricated Si_3N_4-based materials from pure α-Si_3N_4 and MgO of which two compositions (A and B) are given in Table 6.1. These compositions are also depicted in Fig. 6.1, in the ternary phase diagram of these components (Fig. 6.2) and in the Si_3N_4-Si_2N_2O-Mg_2SiO_4 compatibility triangle (Fig. 6.3) for which they were initially chosen because the binary and ternary eutectics in this compositional area are known.

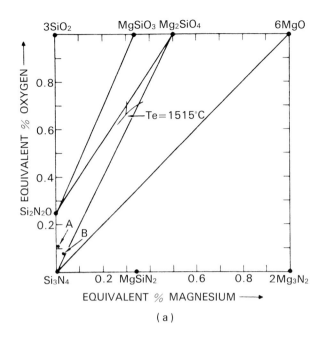

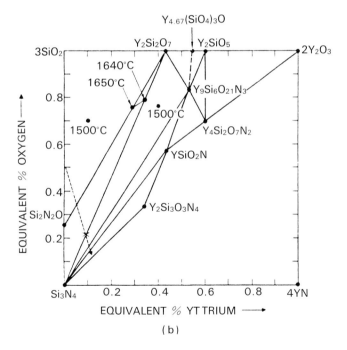

Fig. 6.1 Subsolidus phase relationship in (a) Si-Mg-O-N system (9) and (b) the Si-Y-O-N system (9). See text for discussion of points A and B and the ternary eutectic (T_e) shown in (a).

Table 6.1 Compositions, experimental parameters and results of creep experiments on hot pressed Si_3N_4. (See also Figs. 6.2 and 6.5)

Comp. Symb.	Si_3N_4 (mol %)	SiO_2 (mol %)	MgO (mol %)	Y_2O_3 (mol %)	Other Imp. (mol %)	Temp. Range (K)	Stress Range (MPa)	$\dot{\varepsilon}$ Range (hr^{-1})	atm.	n	ΔH (kJ/mol)	Investigator (Ref.)
A	75.5	22.5	2.0	—	—	1,573–1,673	70–700	6×10^{-5}–10^{-3}	air	~1	660	Lange et al.[11,12]
B	75.5	9.0	15.5	—	—	1,573–1,673	70–700	9×10^{-5}–10^{-2}	air	~2	1080	Lange et al.[11,12]
C	85.5	10.0	—	5.0	—	1,673	150–600	2×10^{-4}–2×10^{-3}	air	1–2	—	Lange et al.[19]
D	92.5	3.0	3.0	—	2.0	1,422–1,588	30–110	10^{-5}–1.5×10^{-3}	air, He	—	546	Kossowsky[28,29]
E	94.0	2.5	2.5	—	1.0	1,422–1,588	30–110	1.6×10^{-7}–10^{-2}	air, He	2.48 3	630	Kossowsky[28,29]
						1,523–1,673	17–220	2×10^{-5}–2×10^{-3}	air N_2 Ar Vacm.	2.0	703	Seltzer[31]
						1,473–1,673	55–139	1.5×10^{-6}–6×10^{-4}	?	1.7	586	Ud Din[35]
F	94.5	2.0	2.5	—	?	1,523–1,673	17–220	2×10^{-6}–2×10^{-4}	air N_2 Ar Vacm.	1.8–2.0 703		Seltzer[31]
G	96.0	2.0	2.0	—	?	1,590–1,873	230–360	7×10^{-8}–2×10^{-7}	?	2.3	650	Wilshire[32–34]
H	93.0	2.0	5.0	—	?	1,590–1,873	230–300	5×10^{-7}–9×10^{-7}	?	2.3	650	Wilshire[32–34]
I	96.0	2.0	—	2.0	?	1,590–1,873	210–300	5×10^{-8}–8×10^{-8}	?	2.1	650	Wilshire[32–34]

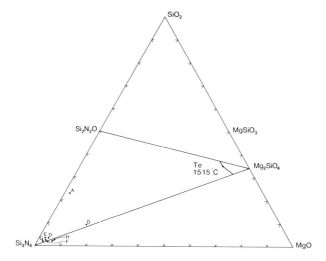

Fig. 6. 2 Si_3N_4-Mg_2SiO_4-SiO_2 compatibility diagram displayed together with compositions used by investigators reviewed in text on the Si_3N_4-SiO_2-MgO phase diagram. Points A and B from Ref. 11); Points D and E from Refs 28), 32) and 35); Point F from Ref. 31); Points G, H and I from Refs. 32-34). Compositions are in mole percent.

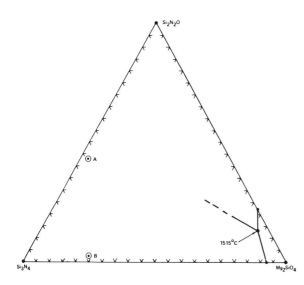

Fig. 6. 3 The Si_3N_4-$MgSiO_4$-Si_2N_2O system showing compositions A and B used by Lange et al. [after F. F. Lange et al.[11]]

Materials A (MgO/SiO_2 molar ratio-0.1) and B (MgO/SiO_2 molar ratio- ~1.6) have considerable compositional differences (in this compatibility triangle) relative to one another and the ternary eutectic composition (*Te*) shown in Figs. 6. 1 to 6. 3. X-ray diffraction revealed that both materials contained β-Si_3N_4 and WC (from the milling process). Composition A also contained Si_3N_4 in apparent proportion to that indicated by its position in the phase diagram. This phase was not detected in composition B. The expected Mg_2SiO_4 was not observed in any composition. The composition of the noncrystalline phase revealed by TEM[17] to be present between all grains and at triple points (see Fig. 6. 4) was assumed to be the same as the ternary eutectic composition for the reasons noted above. (Since no nitrogen was detectable in the noncrystalline phase by EELS or X-ray

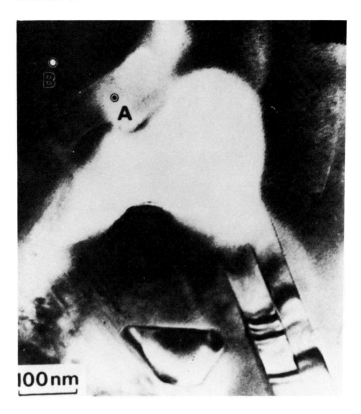

Fig. 6. 4 Transmission electron micrograph showing noncrystalline grain boundary phase in hot pressed Si_3N_4 containing MgO and SiO_2 in amounts similar to composistion B used by Lange et al.[11] and noted in text. [after D. R. Clarke et al.[17]]

microanalyses,[17] this assumption must be considered slightly incorrect. Clarke et al.[17] estimate that the composition actually lies near the SiO_2-$MgSiO_3$ tie line (see Fig. 6. 1).) The volume content of this noncrystalline phase was not qualified; however, composition A being further from the eutectic was expected to contain the least amount of glass. By contrast, composition B, being much closer to the eutectic was expected to contain much more glass.

All creep experiments were performed in air using compressive stress and temperature ranges of 70–700 MN/m^2 and 1,573–1,673 K, respectively. Figure 6. 5 shows that composition B is less creep resistant and has a nominal stress exponent of ~ 2; whereas, composition A has a nominal stress exponent of ~ 1. The activation energy has also been determined[12] in these same materials to be 660 kJ/mol and 1,080 kJ/mol (a very large value indeed) for compositions A and B, respectively, as shown in Figure 6. 6. It should be noted that a true steady-state creep condition was not observed. The total strain did not exceed 0.05 in any sample.

The stress exponent of ~ 1 and the lower activation energy suggest that composition A with the smallest amount of viscous phase creeps by a mechanism controlled by diffusion,

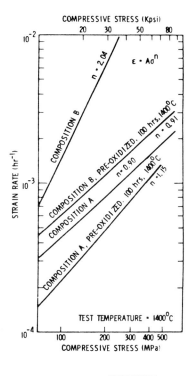

Fig. 6.5 Steady-state compressive creep vs. stress for Si-Mg-O-N materials of composition A and B of Table 6.1. The effect of oxidation induced phase changes on creep behavior in these materials is also illustrated. [after F. F Lange[15)]]

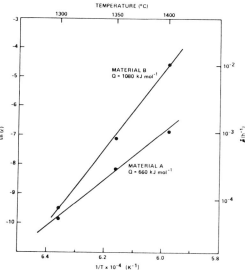

Fig. 6.6 Arrhenius plot of steady-state creep rates for materials A and B of Table 6.1.[after F. F. Lange and B. I. Davis[12)]]

presumably through the viscous grain boundary phase. In contrast, precise density measurements as well as TEM revealed that composition B exhibits extensive cavitation; whereas, relatively few voids were found in composition A. Most cavities were observed at 3 grain junctions; however, a few grains were observed to separate normal to their common boundary, leaving frozen glassy fibrils joining two grains. TEM also revealed the grains to be relatively dislocation free, suggesting that dislocation motion is not a dominant mechanism of creep in Si_3N_4 in the temperature and stress regimes employed in this research.

The preceding information indicates that two different compositionally dependent creep mechanisms can dominate in Si_3N_4 materials fabricated in the same compatibility triangle. The creep behavior of the composition (A) most remote from the ternary eutectic is dominated by an apparent diffusional mechanism; whereas, cavitational creep is most pronounced in the composition (B) closer to this eutectic. Both mechanisms are governed by the viscosity of the noncrystalline phase. Raj[18] has subsequently developed a methodology for separating the creep strain from the cavitational strain and employed this theory with Lange et al.'s data for sample B. These conclusions suggest that the bulk composition should be selected as far as possible from the eutectic composition to minimize the residual glass phase. Unfortunately these compositions are more difficult to densify.

The high temperature creep properties can also be improved by choosing a system in which the additive more easily crystallizes to leave less residual glass. The introduction of Y_2O_3 into Si_3N_4 with $Y_2O_3/SiO_2 \geq 0.55$ (see Fig. 6. 1(a)) produces on heating crystalline Y_2SiO_7 as the major second phase. The creep resistance of this material (comp. C in Table 6. 1) has been found[19] superior to that of the Si_3N_4 materials noted above. The problems of obtaining crystallization of an amorphous grain boundary phase relative to its bulk analog have been discussed by Raj and Lange.[13]

For compositions between A and B, the contribution of cavitation decreases as the composition is shifted toward A. This has been shown[11] by the significant reduction in the apparent steady state creep rate in A and B samples which had been preoxidized (see Fig. 6. 5). The dominant creep mechanism in sample B also changed from cavitation to diffusion. The chemical reasons for this phenomenon have been detailed elsewhere.[20]

Finally it should be noted that cavitation increases with increasing stress. At lower (50-175 MN/m^2) stresses, the strain rates of all the materials are more similar relative to their large divergence at higher (200-700 MN/m^2) stresses. This suggests the dominance of diffusional creep at lower stresses in all materials relative to the dominance of cavitational creep in material (B) prone to cavitate at higher stresses.

With the foregoing background regarding the integrated study of phase equilibria, analytical TEM and creep in Si_3N_4, we are now in a position to review and, to an extent, interpret the results of this research.

The earliest detailed analyses of creep in hot-pressed Si_3N_4 were performed in tension by McLean et al.[25-27] and Kossowsky et al.[28,29] Three stage creep curves similar to those found in metals and alloys were observed. Steady state creep (when observed) accounted for most of the strain while the third stage was rather short and, in many cases, not detected at all. Only the research of the latter investigator will be described in detail.

Kossowsky et al. investigated the creep in two commercial grades of Si_3N_4, HS-110 and HS-130*, the general composition of which are shown as points D and E in Fig. 6. 2. The former material contained $\sim$ 1 wt% Al and $\sim$ 0.5 wt% Ca as compared with 0.1 wt% Al and $\sim$ 0.04 wt% Ca in HS-130. The native SiO_2 content was assumed to be 2-3 wt% based data from Lange.[9] The samples were crept both in air and He under the conditions noted in Table 6. 1. Both the ranges and end point values of temperature and stress were less than those used by Lange et al.; however, the compositions were somewhat similar to

---
* Norton Comp., Worcester, MA. (HS-110 is no longer available.)

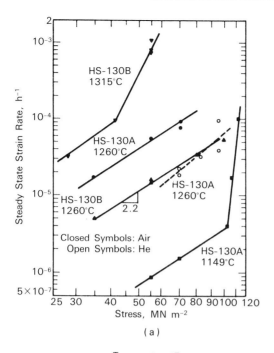

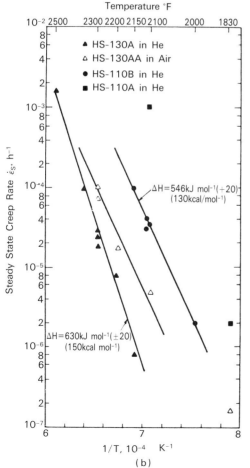

Fig. 6.7 Steady-state creep rates in hot pressed Si_3N_4 (a) as a function of stress, and (b) as a function of temperature. All materials are from the Norton Company, Worchester, MA; however, the HS-110 materials contain more impurities (1 wt% Al and 0.5 wt% (a)) relative to the HS-130 materials (0.1 wt% Al and 0.04 wt% Ca). Type A materials also contain more impuriries (particularly Ca) than type B. [after R. Kossowsky et al.[29]]

composition B with regard to phase relationships and liquid formation.

Specimens crept in air·had a higher creep strain and creep life than those tested in He.[28,29] The creep rate in air for HS-130 was proportional to the stress to the power n where n was ~ 2 between 1,422 K and 1,473 K and 55 and 100 MN/m² as shown in Fig. 6. 7(a). At 1,588 K in air, the value of n remained at ~ 2 up to 40 MN/m²; above this stress an abrupt change in slope occurred to give a value of $n = 4.8$. At 1,533 K in He, n was reported to be 3 indicating an atmospheric effect similar to that noted by Lange et al.[11] An Arrhenius analysis, Fig. 6. 7(b) of the temperature dependence of the secondary creep rate leads to values of the activation energies of 535, 630, and 546 kJ/mole for the HS-130 tested in air, HS-130 in He and HS-110 in He, respectively. Failure was primarily intergranular at all temperatures. The creep rates were also sensitive function of Ca concentration in the grain boundary glas phase, decreasing by some three orders of magnitude as the level of this impurity was decreased.

Kossowsky et al.[28,29] described their creep results in terms of a simple model involving grain boundary sliding, the rate of which was postulated to be controlled by the viscosity of the grain boundary glass phase. This viscosity was, in turn, believed to be controlled primarily by the Ca concentration. This sliding produced cracks at triple points during primary creep. These cracks subsequently grew during secondary creep and joined to produce major cracks which propagated during tertiary creep with subsequent failure.

The occurrence of grain boundary sliding during secondary creep, as required by the above viscous flow model is supported by several observations. Firstly extensive triple point cracking was observed. Secondly there was considerable intergranular cavitation. Finally, the activation energy for secondary creep is in approximate agreement with that for viscous flow in silicate glasses (600 kJ/mole[30]). These results (except activation energy) are similar to those found by Lange et al.[11] for composition B whose creep was dominated by cavitation.

Using the common model for triple point cracking, these authors[28,29] explained the differences in the creep rates in air and in He in terms of capilliary feeding of a surface silicate along the intergranular region. The greater creep life in air was explained in terms of healing of surface cracks by the oxidation products. As noted above, a more detailed explanation has been offered by Clarke and Lange.[20]

Seltzer[31] subsequently conducted both tensile and compression experiments on HS-130 as well as the more impurity-free NC-132 (compositions E and F in Table 6. 1 and Fig. 6. 2) in the temperature and stress ranges of 1,523-1,673 K and 17-220 MN/m² in air, N_2, Ar and vacuum. An activation energy of 703 kJ/mole and stress exponent values of $1.8 < n < 2$ were determined for these two materials. At 1,673 K creep rates were a factor of five greater for tests conducted in air compared with those performed under Ar. Nitrogen had an intermediate effect. The creep behavior was also ascribed to grain boundary sliding controlled by the viscosity of the glass phase on the grain boundaries. As in the studies of Lange et al.[11] and Kossowsky et al.,[29] this sliding was accommodated by void formation, primarily at triple points.

The research of Kossowsky et al.[29] also showed that a compressive stress of 68.9 MN/m² was required to obtain a creep rate of ~ 1.5×10^{-7}/sec at 1,643 K. This was more than ten times the stress needed to achieve this creep rate under tensile creep conditions at the same temperature. Using the reasoning that processes solely dependent on the viscosity of the grain boundary phase should have comparable creep rates in tension and compression, Wilshire and coworkers[32-34] suggested that the rate-determining process during creep is not sliding per se but the accommodation of sliding by grain boundary microcrack development, as noted by Kossowsky et al.[29] and Lange et al.[11] Since in a compression test the maximum tensile stresses generated are only about a tenth of the applied stress, as noted above, and intergranular crack formation depends on the tensile stress level across grain boundaries, the greater strength in compression would therefore be expected.

These authors also conducted creep research on Si_3N_4 materials containing 2 wt% MgO, 5 wt% MgO, and 2 wt% Y_2O_3 in the compressive stress and temperature ranges of 210-360 MN/m² and ~ 1,590-1,873 K. As shown in Fig. 6. 8, the creep strength of the material containing 2% MgO was considerably greater than that for materials of comparable density but containing ~ 5% MgO; however, the Y_2O_3-containing material was superior to both. The stress dependences of the creep rates ($n = 2.1$-2.3) and activation energies (curves not shown) ($Q \sim 650$ (± 25) kJ/mole) were approximately the same for all three materials. These authors argue that while the composition and amount of grain boundary phase affect the creep resistance, the similarity among the n values and among the activation energy values indicates that the operative deformation process is the same in all three materials. Furthermore, the activation energy for viscous flow of silicates varies with composition of

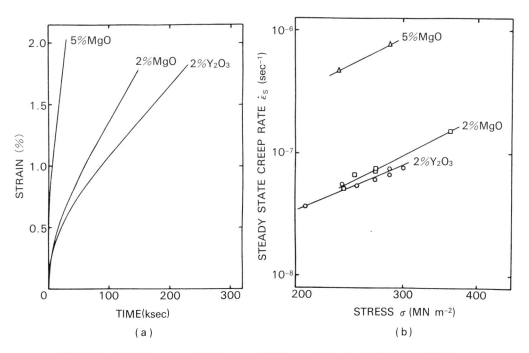

Fig. 6. 8 (a) Creep curves for hot-pressed Si_3N_4 containing 2 wt% MgO, 5 wt% MgO or 2 wt% Y_2O_3 as pressing additives determined at 238 MN/m² and 1,623 K.
(b) Variaton of the steady-state creep rate with applied stress for the same materials noted in (a).
[after P. J. Dixon-Stubbs and B. Wilshire[34]]

the melt. The Q values for creep would therefore be expected to depend on the amount and type of additive if processes such as grain boundary sliding, viscous phase transfer, etc., determine the creep rate. However, the independent value of Q for creep found in this research indicates that the creep is controlled by microcrack formation with the impurity levels acting to modify the creep resistance by affecting the ease of sliding and microcrack development.

In comparing these results and conclusions with those of Kossowsky et al.[29] and Lange et al.,[11,12] one must take note that the stresses used by Wilshire and Lange et al. were more than twice those used by Kossowsky. Also the temperatures used by Kossowsky were less than those of the other investigators. Thus at those higher stress and temperature levels, the void formation and microcracking may have occurred to a greater extent than in Kossowsky's experiments and, therefore, leading to the comments concerning mechanisms noted above.

Ud Din and Nicholson[35] have also investigated the creep behavior in HS-130 Si_3N_4 within the same stress and temperature ranges employed by Kossowsky and Seltzer. However, lower values of the activation energy (586 kJ/mol) and the stress exponent (1.7) were determined. Void formation at triple points as well as limited plastic deformation were observed at the triple points by TEM. These processes were theorized to occur concurrently as a non-Newtonian grain boundary sliding process for creep in this material.

(b) Reaction-Bonded Silicon Nitride (RBSN)

Grathwhol and Thümmler have conducted extensive creep studies in this material as well as authored two reviews[36,37] on this subject. Because of this and the limited space for this article, only a brief discussion of the deformation in this material will be presented.

Initial quantitative creep studies on RBSN have been conducted in air by Mangels,[38] Ud Din and Nicholson,[39] Birch et al.,[32] and Lenoe and Quinn.[40] Creep rates below 10^{-6}/hr at 1,573 K and 70-100 MN/m^2 as well as activation energies between 540 and 700 kJ/mol and stress exponents of $1 < n \leq 2.3$ were reported. These investigators pointed to a creep mechanism of viscous flow of a grain boundary phase controlled by boundary separation.

More recent research by Grathwhol and Thümmler[41,42] has revealed that internal oxidation as a result of oxide introduction during preparation or during creep in air is an important factor in determining the aforementioned creep behavior in this material. (An oxide free RBSN maintained in this condition during the entire creep run (e.g., in vacuum), appears to be practically creep resistant.) The oxygen is introduced via pore channels and grain boundaries. In turn, impurity elements (e.g., Ca, Mg) from the grain boundaries migrate to the surface under oxidizing conditions. This increases the viscosity in the noncrystalline grain boundary phase which causes a decline in the creep and an increase in the activation energy.

Secondary creep in RBSN (when it occurs) does not appear to be a real deformation process, but is accompanied by initiation of cracks and crack growth. A redistribution of micro porosity has been observed and believed to be caused by grain boundary separation during a pure sliding process. Thus if the density of this material could be improved, its practical utility could be greatly enlarged.

6.3 Silicon Carbide

6.3.1 Structure and Polymorphism

The fundamental structural units in SiC are also covalently bonded (sp^3) tetrahedra (either SiC_4 or CSi_4) linked through their corners to satisfy the four-fold coordination at any structure point and to occupy an array of positions analogous to those occupied by spheres in close packed structures. This produces a framework structure of identical polar layers of Si_4C (or C_4Si).

When the same chemical compound exists in two or more crystallographic forms, these forms are called polymorphs and the phenomenon is called polymorphism. In certain close-packed structures such as SiC, there is a special one-dimensional type of polymorphism called polytypism. Polytypes are alike in the two dimensions of the close-packed planes but differ in the stacking sequence in the dimension perpendicular to the close-packed plane. Because of the close-packed structure in SiC, the stacking sequence can be described by the ABC notation. If the pure ABC stacking is repetitive, one obtains the zincblende structure. This is the only cubic SiC polytype and is referred to as β-SiC or the 3C structure (because of the three layer stacking sequence). The hexagonal (...ABAB...) stacking sequence is also found in SiC. In addition, both stacking types can also occur in more complex, intermixed, forms yielding a wider range of ordered, larger period, stacked hexagonal or rhombohedral structures of which 6H is the most common. All of these noncubic structures are known collectively as α-SiC.

6.3.2 Processing Routes

As in the case of silicon nitride, SiC is a generic name for a material which is fabricated by a variety of processing schemes which result in a multitude of microstructures and deformation mechanisms. Sintering or hot pressing without B- or Al-containing or oxide additives or the use of extreme pressures appears impossible; thus, reaction bonding and chemical vapor deposition (CVD) are also common processing techniques. All of these process routes produce materials which are essentially or fully dense.

Reaction-bonded SiC is produced by slip casting shapes from a slurry of colloidal graphite and Acheson-derived SiC, with the latter having a bimodal grain-size distribution (~ 100 μm and ≤ 10 μm). The plate is then heated in a Si-containing N_2 atmosphere at $T > 2,273$ K to obtain reaction with the free C and resultant bonding. Free Si results as a nongrain boundary second phase which varies in amount throughout each sintered piece.

In the CVD process, several combinations of source gases and procedures may be used; however, the most common commercial scenario is the pyrolysis of methyltrichlorosilane ($CH_3SiCl_3 \rightarrow SiC + 3HCl$) on graphite at $\simeq 1,673$ K. Because of inherent residual stresses following growth, the material is normally annealed in purified Ar at 10^7 N/m^2 and 2,373 K. The substrate is subsequently ground away.

A neo-classical method of producing sintered SiC has been hot pressing a mixture of Acheson-derived grains and Al_2O_3 or Fe_2O_3 at temperatures sufficient to form a liquid phase at the grain boundaries between the SiO_2 on the SiC and the oxide additive. For SiC materials which must be used at high temperatures, this process has been largely replaced by more recently developed sintering or hot pressing technologies wherein small (one weight percent or less) amounts of B- or Al-containing substances (usually the carbides or nitrides) are mixed with very fine SiC powders, molded into the desired shape and fired in an inert atmosphere to 2,400–2,500 K. Extensive TEM[45,46] and Auger[45,46] analyses have not revealed B on the grain boundaries. By contrast, Al is always found on the boundaries. These results may be indicative of two different mechanisms of densification.

The above procedures are those employed in producing the samples on which were conducted the creep studies described in the following paragraphs.

6.3.3 Deformation and Creep
(a) Reaction-Bonded SiC

In previous creep research on reaction-bonded SiC, Rumsey and Roberts[47] used four-point bending at 1,473 K and 141 MN/m^2. They reported a steady-state creep region wherein the creep rate was approximately 2.7×10^{-10} s^{-1}. The onset of creep was attributed to the deformation of the free Si on the SiC grain boundaries. Subsequent and similar research by marchall and Jones[48] revealed only transient creep behavior, even after 3.6 Ms at $T = 1,273$ to 1,473 K and stresses from 207 to 414 MN/m^2. Although an activation energy of 230 ± 79 kJ/mol and stress exponents which varied from one to two with increasing temperature were calculated, no conclusion concerning the controlling mechanism could be reached.

Four-point bending experiments were also conducted by Larsen and coworkers[49,50] on four reaction-bonded materials: NC-435,*1 Refel,*2 NC-430,*3 and Coors Si/SiC.*4 All of the reported data were derived from tests at 1,623 K and stresses from 34 to 206 MN/m^2. The NC-430 had the best creep resistance (creep rate (3.0 to 10.0) $\times 10^{-10}$ s^{-1}), which was approximately an order of magnitude lower than the creep rates of the other materials. All of the materials had a stress exponent of $\simeq 1$ with Coors having the highest at 1.7 and NC-435 the lowest at 0.9. (The stress exponent for NC-430 was 1.2.) They reported much scatter of their data, thus making the stress-exponent determination difficult. This scatter was attributed to a variation of the amount of free Si in the material. Although no activation energies were calculated, diffusion was suggested as the rate-controlling process.

Krishnamachari and Notis[51] reported an activation energy for KT SiC of 146 kJ/mol using temperature and stress ranges of 1,573 to 1,673 K and 34 to 86 MN/m^2, respectively. The mechanism of creep was concluded to be Coble grain-boundary diffusion with Si being the rate-controlling species. This was in agreement with their deformation map for polycrystalline SiC having a grain size of 65 μm.

In a communication with Thummler of Karlsruhe, West Germany, it was learned that Schnürer et al.[52] have conducted creep experiments in vacuum and in air on reaction-bonded SiC in the bending mode at 1,273 to 1,673 K. The stress exponent increased from

---

*1 Norton Co., Worcester, MA.
*2 United Kingdom Atomic Energy Authority/British Nuclear Forum, Springfields, UK.
*3 Norton Co.
*4 Coors Procelain Co., Golden, Co.

one to four as the stress was raised from 97 MN/m² (14,000 psi) to 190 MN/m² (27,500 psi). This increase in the stress exponent was originally attributed to the increased mobility of dislocations with an increase in applied stress but has since been attributed to free Si at the grain boundaries.[53]

Finally, Seltzer[54] performed compressive creep tests on NC-435 SiC. A stress exponent of ∼ 1 was obtained from the small amount of data taken at 1,673 K and stresses of 110 and 138 MN/m². The creep rate was on the order of 5.5 to 8.3 × 10⁻¹⁰ s⁻¹.

Although several of the previously mentioned researchers believe that creep in reaction-bonded SiC is caused by the Si phase at the grain boundaries, electron microprobe analysis by Marchall[55] and transmission electron microscopy by the present authors (see discussion below) revealed no grain-boundary second phase.

Another point worthy of note is that all of the above investigators reported deformation of reaction-bonded SiC at T≤1,673 K and stresses comparable to those used in the research recently conducted by the present authors and reported below. However, we measured no deformation at or below 1,848 K and 69 MN/m² (10,000 psi). Except for the work by Seltzer,[54] all of the previous experiments were conducted in four-point bending. It is believed that some, if not all, of the deformation was the result of slow crack growth such as that noted by McHenry and Tressler[56] in hot-pressed and sintered SiC stressed in the bending mode below 1,673 K. However, the available data are insufficient to confirm this point.

As previously indicated, Carter et al.[57] have also conducted creep research on a reaction-bonded SiC (NC-430) in the temperature and constant compressive stress ranges of 1,848-1,923 K (melting point of Si = 1,693 K) and 110-220 MN/m², respectively. In the determination of the stress exponent, the plot of the data for $\dot{\varepsilon}$ as a function of σ/G showed considerable scatter. However, a plot of the raw data as a function of the density of the uncrept sample showed that much of the scatter corresponds to density variations from sample-to-sample in the as-received material. The range of densities in the as-received samples was caused by the 1.3 vol% variation in the free Si originally incorporated in the material.

To compensate for the differences in density between samples, two mathematical procedures were used by Carter et al.[57] Initially, a linear regression was performed on the plot of the values of the log of the steady-state strain rate vs. the measured density values. The slope and the intercept of the resulting linear curve were found to be -0.073 and 218.8, respectively. The compensated strain rate for a given sample was calculated by (1) computing the strain rate for the mid-range value of density (3.108×10^3 kg/m³) using the equation (log $\dot{\varepsilon}$ = 218.8 + 0.073 (ρ)) from the above regression, (2) taking the ratio of this value to the strain-rate value calculated from this last equation using the measured bulk density of the sample, and (3) multiplying this ratio by the sample strain rate taken from the raw data. The log of this density-compensated steady-state strain rate was then plotted as a function of log σ/G, as shown in Fig. 6.9. The stress exponent was calculated from this compensated data to be 5.7.

The activation energy for creep of this material, determined from the standard plot of $\dot{\varepsilon}$ vs. $1/T$ shown in Fig. 6.10, was calculated to be 711 ± 21 kJ/mol from data collected

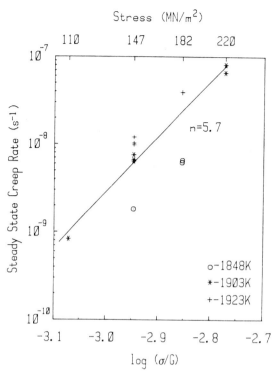

Fig. 6.9 Steady-state creep rate (compensated for density) vs. log (σ/G) for all NC-430 SiC samples. [after C. H. Carter, Jr. et al.[57]]

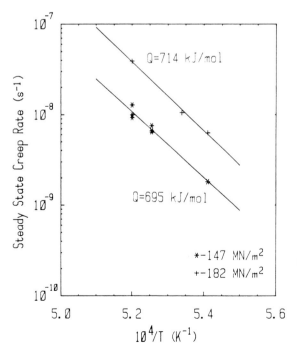

Fig. 6.10 Steady-state creep rate (compensated for density) vs. $10^4/T$ for NC-430 SiC. [after C. H. Carter, Jr. et al.[57]]

at 147 MN/m² and 183 MN/m².

The as-received NC-430 SiC contained a large number of dislocations as well as low angle boundaries, as can be seen in Fig. 6. 11. These boundaries are the interfaces between the Acheson-derived SiC and the SiC that was formed during reaction-bonding. This was determined by observing the boundaries at low magnification and noting that they are approximately parallel to the edge of the grains and are at a distance from the edge that is approximately the thickness of the newly formed SiC. This thickness was determined from impurity-sensitive, secondary-electron-imaged SEM micrographs of polished, unetched material. The material on both sides of these boundaries was determined to be α-SiC. In addition, these boundaries were distinguishable from some other low-angle boundaries (to be shown later) by the cavities that lie along them.

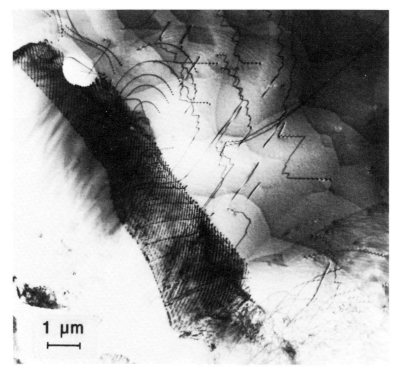

Fig. 6. 11 High voltage transmission electron micrograph of as-received NC-430 SiC showing low angle boundary, cavity on boundary and dislocation source.[after C. H. Carter, Jr. et al.[57]]

The slip traces seen in Figs. 6. 12(a) and (b) indicate that considerable dislocation glide is occurring. Other interesting features in Fig. 6. 12(a) are the kinks (or jogs?) and the pile-up of the slip bands at some type of low angle boundary.

Burgers vector analysis of the dislocations in the slip bands of Fig. 6. 12(b) revealed that these dislocations are Shockley partials in the hexagonal system having the vector $a/3<01\bar{1}0>$.

A comparison of the stress exponent value with that of theoretical models indicates

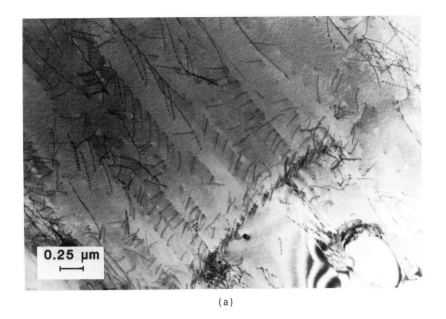

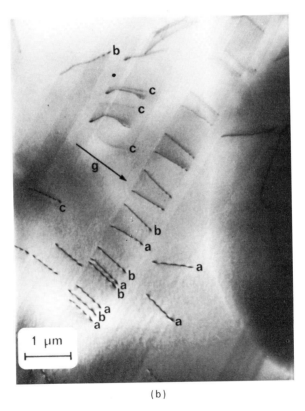

Fig. 6.12 High-voltage transmission electron micrograph of NC-430 after creep. (a) Slip bands of which the majority end at a type of low angle boundary. (b) Slip band on which Burgers vector analysis was performed. Those dislocations labeled a are $\vec{b} = a/3[01\bar{1}0]$; b are $\vec{b} = a/3[10\bar{1}0]$, and c are $\vec{b} = a/3[1\bar{1}00]$. [after C. H. Carter, Jr. et al.[57]]

that dislocation climb by pipe diffusion is the controlling mechanism; however, the similarity in values of the activation energies for lattice diffusion of Si in α-SiC (789 ± 10 kJ/mol)[58] and for creep in α-SiC (the only form determined to be present in the creep samples) indicates that dislocation glide/climb controlled by climb is the controlling process.

The TEM studies definitely indicate a dislocation mechanism. One of the most notable differences in the as-received and crept material in the TEM is the appearance of slip bands such as those seen in Figs. 6. 12(a) and (b). These slip bands are proof that there is considerable dislocation movement by glide occurring as a result of creep.

The primary indicator that climb also occurred in this material was the observation of the formation of low-angle boundaries (or cell walls) during the creep process, as shown by the horizontal boundary in Fig. 6. 13. These boundaries can be formed only if either (1) climb is occurring or (2) there are enough slip systems and the resolved shear stresses are proper to achieve cross slip. There are two major reasons which make the former process much more likely. The first is that all of the dislocation bands were observed to lie in the basal plane. Therefore, not enough operative slip systems to allow cross slip were observed in this material at the temperatures and stresses used. Second, the cell wall contained many dislocation loops. These loops would occur in a low-angle boundary such as this only if climb were active.

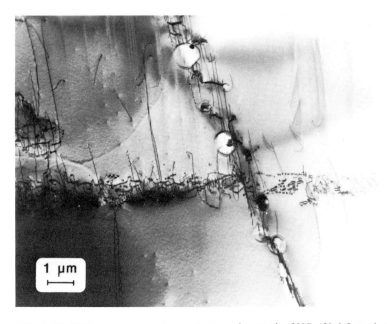

Fig. 6. 13 High-voltage transmission electron micrograph of NC-430 deformed at 1,903 K and 147 MN/m² showing low angle boundaries; some do not contain cavities and lie at high angles to those containing cavities. [after C. H. Carter, Jr. et al.[57]]

(b) Chemically Vapor Deposited SiC

The only deformation research on CVD SiC is that of the present writers[59] who employed temperatures and constant compressive stresses from 1,848 to 2,023 K and 110 to 220 MN/m², respectively. The stress exponent was found to vary from 2.3 at 1,823 K to 3.7 at 1,923 K as shown in Figs. 6. 14(a) and (b). This data has very little scatter compared to that of the raw data from NC-430 SiC; since, there was very little variation in the density of the samples. The average value of the activation energy was calculated to be 174.6 ± 4.8 kJ/mol from the graphs shown in Figs. 6. 15(a) and (b).

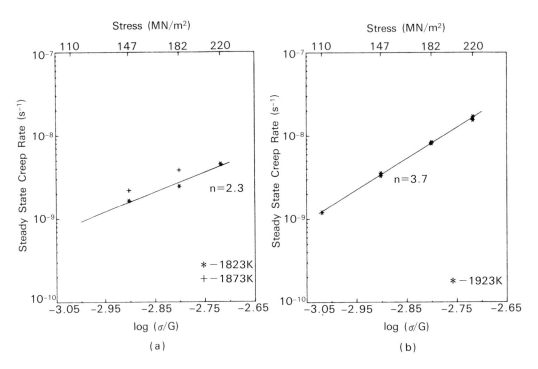

Fig. 6. 14 Steady-state creep rate vs log (σ/G) for CVD SiC samples crept at temperatures (a) less than 1,923 K and (b) at 1,923 K. [after C. H. Carter, Jr. et al.[59]]

Complete faulting existed in the as-deposited material as shown by the very high density of lattice fringes in Fig. 6. 16(a). The stacking sequence of the basal plane in the as-deposited material was almost totally random as was evident by the very considerable streaking in the diffraction pattern. Annealing allowed exaggerated grain growth caused by the movement of high angle boundaries on grains oriented off the preferred orientation. This step also produced a more ordered stacking sequence of the basal planes, particularly in grains experiencing exaggerated grain growth.

It was determined that the primary α polytype in the CVD material is 6H. Furthermore, there was roughly a 60 : 40 ratio of β to α which existed in the as-deposited material and which was maintained in the annealed and crept samples. There was, however, a redistribution of these polytypes during annealing, as shown in Fig. 6. 16(b).

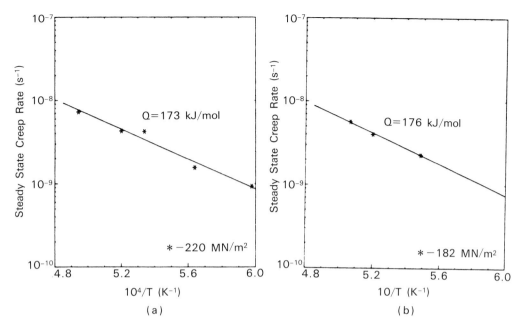

Fig. 6. 15 Steady-state creep rate vs $10^4/T$ for CVD SiC samples crept at (a) 182 MN/m² and (b) 220 MN/m². [after C. H. Carter, Jr. et al.[59]]

The dislocations in the as-received material were determined to be primarily of the type $b = a/b<114>$ or $b = a/2<\bar{1}10>$, both of which are produced by the combination of a Shockley partial and a Frank partial and are sessile. All dislocations were found to lie on the (111) plane.*

Slip bands were the predominant dislocation structure observed in the as-crept material at all stresses and temperatures (see Fig. 6. 17); however, they were never seen in the as-received material. As in the NC-430 SiC, these dislocations are Shockley partials which move in alternating pairs. The specific Burgers vectors seen in the slip bands shown here are $a/6\ [2\bar{1}\bar{1}]$ and $a/6\ [1\bar{2}1]$.

The calculated value of the activation energy for steady-state creep in this material of 175 kJ/mol is much lower than the activation energy values for self-diffusion of Si or C in α- or β-SiC.[58,60-62] Even if the lowest value of 563 kJ/mol which is that for boundary diffusion of C in β-SiC is taken for comparison, this latter value is still 3.2 times higher than the value of the activation energy for creep in this material. On this basis it is deemed unlikely that the controlling creep mechanism is diffusion-controlled.

The range of values for the stress exponent indicates that the controlling creep mechanism is either dislocation motion or grain boundary sliding. The results of the TEM study prove that the creep mechanism definitely involves dislocation glide. Furthermore, no indication of grain-grain separation at triple points or movement of grains relative to one

---

* The Burgers vectors and diffraction vectors are given as cubic for this material; since, the cubic β-SiC is the dominant phase.

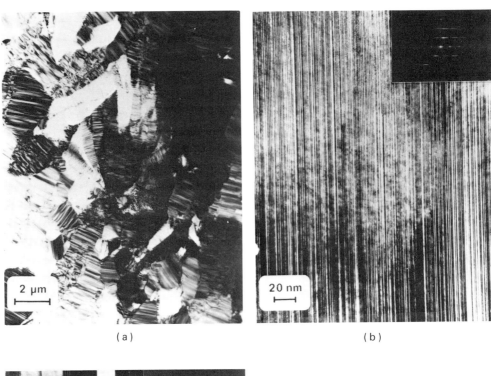
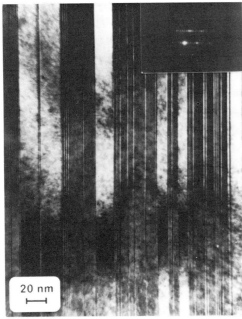

Fig. 6.16 TEM micrographs of CVD SiC. (a) and (b) show heavy faulting of as-deposited material. (c) Lattice image of anneled material deformed at 1,973 K and 1,982 MN/m². The dark areas in (c) are α-SiC and the light areas are β-SiC. [after C. H. Carter, Jr. et al.[59)]]

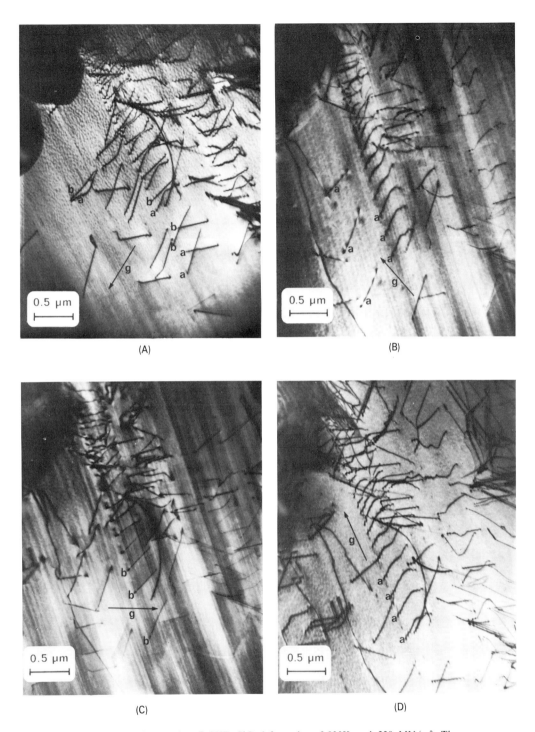

Fig. 6.17 TEM micrographs of CVD SiC deformed at 2,023K and 220 MN/m². The Burgers vectors of the labeled dislocations are as follows: (a) $\vec{b} = a/6[2\bar{1}\bar{1}]$, (b) $\vec{b} = a/6[1\bar{2}1]$. (A) $\vec{g} = [\bar{2}20]$, $z \simeq [112]$, (B) $g = [311]$, $z \simeq [114]$; (C) $\vec{g} = [13\bar{1}]$, $z \simeq [114]$, (D) $\vec{g} = [20\bar{2}]$, $z \simeq [111]$. [after C. H. Carter, Jr. et al.[59)]]

another was evident in any of the large number of micrographs examined. It was therefore concluded that below 1,923 K, the principal creep mechanism is dislocation glide controlled by the energy of the Peiérls stress hills.

Employing an equation developed by Seeger[63] which estimates the activation energy for creep via a double kink mechanism (i.e., a double kink is formed in a dislocation acting against the Peierls stress), it was found that the experimental value roughly approximates the calculated value determined for the CVD SiC.

At high temperatures (>1,923 K), the stress exponent increases to 3.7 and the threshold stress decreases to less than 110 MN/m². The increase in stress exponent is an indication that the controlling creep mechanism may be changing at high temperatures to one involving dislocation glide/climb controlled by climb.

(c) Hot Pressed or Sintered SiC

One of the earliest deformation studies on high density (>99%) polycrystalline α-SiC produced by hot pressing at 2,873 K was conducted by Farnsworth and Coble.[64] It was tentatively concluded that creep in this material in the temperature and stress ranges of 2,173-2,473 K and 20-200 MN/m², respectively, was controlled by diffusion in the grain boundaries. Subsequent work by Francis and Coble[65] on similar hot pressed material but containing varying amounts of porosity and grain sizes provided qualitative support for the aforementioned mechanism. The controlling species was not positively identified; however, carbon was suggested, as the activation energy for lattice diffusion calculated from the creep data (306 ± 63 kJ/mol) was almost half the only experimentally determined value for self-diffusion of C in SiC (586 kJ/mol)[66] available at that time. No stress exponent was determined; however, the material was reported to deform "viscously" which would normally have an n value of approximately one. In additional research,[65] single crystals of α-SiC were stressed in bending to 276 MN/m² at 2,400 K for 3.6×10^4 s. Transmission electron microscopy confirmed the generation and reluctant movement of a few dislocations on the nonbasal $\{4\bar{4}01\}$ and $\{1\bar{1}01\}$ planes; similar dislocation effects on the $\{0001\}$ planes were not observed. Thus, it was concluded that dislocation movement was not important for deformation in poly-crystalline α-SiC. This is the same reasoning reported earlier by Hasselman and Batha[67] who found no plastic flow in α-SiC single crystals subjected to four-point bending up to 2,473 K.

The mechanistic explanation just described for the polycrystalline material has been challenged, however, by Shaffer and Jun[68] who argued that the 1.65% Al (m.p. = 933 K) and 0.97% Fe (m.p. = 1,808 K) reported by Farnsworth and Coble are both in excess of their solubility limits in α-SiC and therefore form a soft metallic liquid grain boundary phase which controls the observed creep behavior in the temperature range of investigation. Utilizing several commercial grades of SiC and a much higher purity material prepared by sublimation, Shaffer and Jun have indeed shown that Young's Modulus for the pure material does not decrease up to 1,743 K in contrast to the impure materials. Unfortunately, the temperatures employed in these investigations were below the temperatures for rapid diffusion and those utilized by the Coble group.

In an additional contribution to the above argument, Seltzer[31] has reported, from very limited data, that the analysis of creep in hot pressed SiC containing a small amount of impurities (principally Al_2O_3) on the grain boundaries resulted in a stress exponent (n) value of 0.88 which could be associated with a boundary mechanism involving a second phase.

Constant load, compressive stress creep research on high density sintered and hot pressed alpha silicon carbides prepared in France and the U.S.A. and containing 1% B and 1% C (sintered material) or 1% B_4C (hot pressed material) has been reported by Djemel et al.[69] The porosity of the material was ~4% with an average grain size of 3.5μm. The applied stress and temperature ranges were 500-1,695 MN/m² and 1,573-1,773 K, respectively. A plot of the resulting strain rates as a function of stress for different temperatures is shown in Fig. 6.18. Two very different regimes of creep are obvious from this graph. These researchers found that below a critical stress, σ_0, the activation energy and stress exponent were 292 kJ/mol and ~1.0, respectively, which is similar to that noted by Coble and coworkers.[64,65] The controlling mechanism was not determined; however, it was postulated to be Coble creep. Although it was not accounted for, variation in porosity also affected the strain rate. In addition, no apparent difference was observed in the microstructure (the porosity was located at triple points except in the sintered material where it was also observed inside the grains) before and after deformation in the first creep regime.

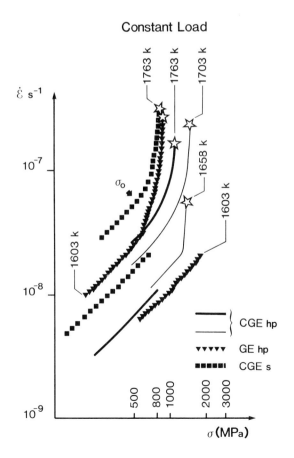

Fig. 6.18 Curves of strain-rate as a function of stress for different α-SiC materials crept at 1,603K, 1,658K, 1,703K, and 1,763K. The CGE hot pressed (hp) and sintered(s) materials were produced by Compagnie Generale d'Electricite in France; the GE (hp) material was produced by the General Electric Company in the U.S.A. [after A. Djemel et al.[69]]

Above σ_0, the strain rate became very high, leading ultimately to fracture as a result of vacancy coalescence along the grain boundaries in tension (i.e., facets parallel to the applied stress) and subsequent cavitation which promoted crack initiation and growth. Thus, diffusion was postulated to control the deformation in both regimes via movement of material below σ_0 and movement of vacancies along the grain boundaries above this stress level.

Finally, Carry and Mocellin[70] have investigated the creep in hot pressed or hot isostatically pressed Al-doped α-SiC in compression (52-110 MN/m^2) between 2,073 K and 2,333 K in vacuum. The deformation was nonreproducible within a given sample type because of the variation in Al content from sample to sample. The controlling mechanism(s) were not determined.

6. 4 Conclusions

Although Si_3N_4 and SiC may be fabricated into bulk material by similar processing routes, their different resulting microstructure and crystal structures cause the kinetics and mechanisms of steady state creep to also be different. The deformation in reaction-bonded Si_3N_4 occurs primarily as a result of the occurrence or formation of second phase oxides. On the other hand, the controlling mechanism in reaction-bonded SiC is dislocation climb within the concurrent mechanisms of glide and climb. The creep of hot pressed Si_3N_4 containing MgO or Y_2O_3 as sintering aids is a sensitive function of the cogent phase equilibria which is extant between the Si_3N_4 and the additive. If the reaction produces considerable amounts of noncrystalline phase, the creep is controlled by cavitation. However, if the content of this phase is small the controlling mechanism becomes that of a solution-precipitation process. By contrast, creep in sintered SiC containing B or Al compounds and C appears to be controlled by lattice and/or grain boundary diffusion.

Acknowledgements

This work has been supported by the National Science Foundation under Grant No. DMR-8022197, the Army Research Office under Contract No. DAAG29-79-G-006 and by the Basic Energy Sciences Division, U.S. Department of Energy through the SHaRE program under Contract No.EY-76-C-05-0033 with Oak Ridge Associated Universities.

References

1) S. Wild, P. Grievson and K. H. Jack : Special Ceramics 5, ed. by P. Popper, Brit. Ceram. Res. Assoc., Stoke-on-Trent (1972) 395-395.
2) I. Kohatsu and J. W. McCauley : "Re-examination of the crystal structure of α-Si_3N_4," Mater. Res. Bull., **9** (1974) 917-920.
3) K. Kato, Z. Inoue, K. Kijima, I. Kawada, H. Tonaka and T. Yamae : "Structural approach to the problem of oxygen content in α-Si_3N_4," J. Am. Ceram. Soc., **58** (1975) 90-91.
4) H. F. Priest, F. C. Burns, G. L. Priest and E. C. Skaar : "Oxygen content of α-Si_3N_4," J. Am. Ceram. Soc., **56** (1973) 395.
5) A. J. Edwards, D. P. Elias, M. W. Lindley, A. Atkinson and A. J. Moulson : "Oxygen content of reaction-bonded α-Si_3N_4," J. Mater. Sci., **9** (1974) 516-517.
6) D. Compos-Loriz and F. L. Riley : "Effects of

6) silica on the nitridation of Si," J. Mater. Sci., **11** (1976) 195-198.

7) J. L. Iskoe, F. F. Lange and E. S. Diaz: "Effect of selected impurities on the high temperature mechanical properties of hot-pressed silicon nitride," J. Mater. Sci., **11** (1976) 908-912.

8) F. F. Lange, S. C. Singhal and R. C. Kuznicki: "Phase relations and stability studies in the Si_3N_4-SiO_2-Y_2O_3 pseudoternary system," J. Am. Ceram. Soc., **60** (1977) 249-252.

9) F. F. Lange: "Phase relations in the system Si_3N_4-SiO_2-MgO and their interrelation with strength and oxidation," J. Am. Ceram. Soc., **61** (1978) 53-56.

10) F. F. Lange: "Eutectic studies in the system Si_3N_4-Si_2N_2O-Mg_2SiO_4," J. Am. Ceram. Soc., **62** (1979) 617; Corr. J. Am. Ceram. Soc., **63** (1980) 231.

11) F. F. Lange, B. I. Davis and D. R. Clarke: "Compressive creep of Si_3N_4/MgO alloys: Part 1 — Effect of composition; Part 2 — source of visco-elastic effect; Part 3 — Effects of oxidation induced compositional change," J. Mater. Sci., **15** (1980) 601-610; 611-615; 616-618.

12) F. F. Lange and B. I. Davis: "Compressive creep of Si_3N_4/MgO alloys," J. Mater. Sci., **17** (1982) 3637-3640.

13) R. Raj and F. F. Lange: "Crystallization of small quantities of glass (or a liquid) segregated in grain boudaries," Acta. Met., **29** (1981) 1993-2000.

14) F. F. Lange: "Fabrication and properties of dense polyphase silicon nitride," Bull. Am. Ceram. Soc., **62** (1983) 1369-1374.

15) F. F. Lange: "High tempereture deformation and fracture phenomena of polyphase silicon nitride materials," in Prog. Nitrogen Ceram., NATO ASI Ser., Ser. E, **65** (1983) 467-490.

16) F. F. Lange: "Silicon nitride polyphase systems: Fabrication, microstructure and properties," Int. Metals Rev., No. 1 (1980) 1-20.

17) D. R. Clarke, N. J. Zaluzec and R. W. Carpenter: "The intergranular phase in hot-pressed Si_3N_4 I, Elemental composition," J. Am. Ceram. Soc., **64** (1981) 601-607.

18) R. Raj: "Separation of cavitation-strain and creep-strain during deformation," Commun. Am. Ceram. Soc. (March 1982) C-46.

19) F. F. Lange, B. I. Davis and H. C. Graham: "Compressive creep and oxidation resistance of an Si_3N_4 materials fabricated in the system Si_3N_4-Si_2N_2O-$Y_2Si_2O_7$," Commun. Am. Ceram. Soc. (June 1983) C-98-99.

20) D. R. Clarke and F. F. Lange: "Oxidation of Si_3N_4 alloys: Relationship to phase equilibria in the Si_3N_4-SiO_2-MgO system," J. Am. Ceram. Soc., **63** (1980) 586-588.

21) B. D. Powell and P. Drew: "The identification of a grain-boundary phase in hot-pressed silicon nitride by Auger electron spectroscopy," J. Mater. Sci., **9** (1974) 1867-1870.

22) D. R. Clarke and G. Thomas: "Grain boundary phases in a hot-pressed MgO fluxed silicon nitride," J. Am. Ceram. Soc., **60** (1977) 491-495.

23) K. V. Lou, T. E. Mitchell and A. H. Heuer: "Discussion of grain boundary phases in a hot-pressed MgO fluxed silicon nitride," J. Am. Ceram. Soc., **61** (1978) 462-464.

24) D. R. Clarke and G. Thomas: "Reply to comments in reference 2," J. Am. Ceram. Soc., **61** (1978) 464.

25) A. F. McLean, E. A. Fisher and R. J. Bratton: Interim Report #AMMRC-CTR 72-19 on Contract DAAG 46-71-C-0162 (Sept. 1973).

26) A. F. McLean, E. A. Fisher and R. J. Bratton: Inrerim Report #AMMRC-CTR 73-32 on ARPA Contract DAAG 46-71-C-0162 (Sept. 1973).

27) A. F. McLean, E. A. Fisher and R. J. Bratton: Inrterim Report #AMMRC-CTR 73-9 on Contract DAAG 46-71-C-0162 (March 1973).

28) R. Kossowsky: "Creep and fatigue of Si_3N_4 as related to microstructures," in Proc. Second Army Mater. Tech. Conf., Ceramics for High Performance Applications, ed. by J. J. Burke, A. E. Gorum and R. N. Katz, Brook-Hill Pub. Co., Cestnut Hill, MA (1974) 347-371.

29) R. Kossowsky, D. G. Miller and E. S. Diaz: "Tensile and creep strengths of hot-pressed Si_3N_4," J. Mater. Sci., **10** (1975) 983-997.

30) R. Rosen, J. Berson and G. Urbain: "Viscosities of silicate melts," Rev. Hautes. Temp. Bt. Refract. (French), **1** (1964) 159.

31) M. Seltzer: "High temperature creep of silicon-based compounds." Ceram. Bull., **56** (1977) 418-423.

32) J. M. Birch, B. Wilshire and D. J. Godfrey:

"Deformation and fracture processes during creep of reaciton bonded and hot-pressed silicon nitride," Proc. Brit. Ceram. Soc., **26** (1978) 141-154.

33) J. M. Birch and B. Wilshire : "The compression creep behavior of silicon nitride ceramics," J. Mater. Sci., **13** (1978) 2627-2636.

34) P. J. Dixon-Stubbs and B. Wilshire : "Creep of hot-pressed silicon nitride ceramics," J. Mater. Sci., **14** (1979) 2773-2774.

35) S. U. Ud Din and P. S. Nicholson : "Creep of hot-pressed silicon nitride," J. Mater. Sci., **10** (1975) 1375-1380.

36) G. Grathwol and F. Thümmler : "Creep of reaction-bonded silicon nitride," J. Mater. Sci., **13** (1978) 1177-1186.

37) F. Thümmler and G. Grathwol : "Creep and internal oxidation of reaction-bonded silicon nitride," in Prog. Nitrogen Ceram., NATO ASI Ser., Ser. E, **65** (1983) 547-555.

38) J. A. Mangels : Ceramics for High Performance Applications, ed. by J. J. Burke, R. N. Gorum and A. E. Katz, Brook-Hill Pub. Co., Chestnut Hill, MA (1974) 195-206.

39) S. Ud Din and P. S. Nicholson : "Creep deformation of reaction-sintered silicon nitrides," J. Am. Ceram. Soc., **58** (1975) 500-502.

40) E. M. Lenoe and G. D. Quinn : Deformation of Ceramic Materials, ed. by R. C. Bradt and R. E. Tressler, Plenum Publishing Corp., New York (1975) 399-404.

41) F. Thümmler, F. Portz, G. Grathwol and W. Engel : Science of Ceramics 8, British Ceramic Soc., Stoke-on-Trent (1976) 133-142.

42) F. Thümmler and G. Grathwol : AGARD Report No. 651 (1976).

43) Y. Tajima and W. D. Kingery : "Grain boundary segregation in aluminum-doped silicon carbide," J. Mater. Sci., **17** (1982) 2289-2297

44) R. Hamminger, G. Grathwol and F. Thümmler : "Microanalytical investigation of sintered silicon carbide, Part 1: Bulk materials and inclusions," J. Mater. Sci., **18** (1983) 353-364.

45) R. Hamminger, G. Grathwol and F. Thümmler : "Microanalytical investigation of sintered SiC, Part 2: Study of the grain boudaries of sintered SiC by high resolution Auger electron spectroscopy," J. Mater. Sci., **18** (1983) 3154-3160.

46) R. F. Davis, J. E. Lane, C. H. Carter, Jr., J. Bentley, W. H. Wadlin, D. P. Griffis, R. W. Linton and K. L. More : "Microanalytical and microstructural analyses of boron and aluminum regions in sintered alpha silicon carbide," Scann. Electr, Micros. (1984) 1161-1167.

47) J. C. V. Rumsey and A. L. Roberts : "Delayed fracture and creep in silicon carbide," Proc. Br. Ceram. Soc., **2** (1967) 233-239.

48) P. Marchall and R. B. Jones : "Creep of silicon carbide," Powder Metall., **12** (1969) 193-201.

49) D. C. Larsen and G. C. Walther : "Property screening and evaluation of ceramic vane materials," AFML Rept. No. IITRI-D6114-ITR-24 (Oct. 1977).

50) D. C. Larsen and J. W. Adams : "Property screening and evaluation of ceramic turbine materials," Rept. No. 11, Contract F33615-79-C-5100 (Nov. 1981).

51) V. Krishnamachari and M. R. Notis : "Interpretation of high temperature creep of SiC by deformation mapping techniques," Mater. Sci. Eng., **27** (1977) 83-84.

52) K. Schnürer, F. Thümmler and G. Grathwol : Universtät Karlsruhe; personal communication (June 1979).

53) K. Schnürer, G. Grathwol and F. Thümmler : "Kriechverhalten Verschneider SiC-Werstaff (German)," in Science of Ceramics 10, Deutsche Keramische Gesellschaft, Berlin (1980) 645-652.

54) M. S. Seltzer : "High temperature creep of ceramics," AFML-TR-76-97 (June 1976).

55) P. Marchall : "The relationship between delayed fracture, creep and texture in silicon carbide," Special Ceramics Symp., Stoke-on-Trent (1967).

56) K. D. McHenry and R. E. Tressler : "Fracture toughness and high temperature slow crack growth in SiC," J. Am. Ceram. Soc., **64** (1980) 152-156.

57) C. H. Carter, Jr., R. F. Davis and J. Bentley : "Kinetics and mechanisms of high temperature creep in silicon carbide: I, Reaction-bonded," J. Am. Ceram. Soc., **67** (1984) 409-417.

58) J. D. Hong and R. F. Davis and D. E. Newbury : "Self-diffusion of silicon-30 in

α-SiC single crystals," J. Mater. Sci., **16** (1981) 2485-2492.

59) C. H. Carter, Jr., R. F. Davis and J. Bentley : "Kinetics and mechanisms of high temperature creep in silicon carbide: II, Chemically vapor deposited," J. Am. Ceram. Soc., **67** (1984) 732-740.

60) J. D. Hong and R. F. Davis : "Self-diffusion of carbon-14 in high purity and N-doped α-SiC single crystals," J. Am. Ceram. Soc., **63** (1980) 546-552.

61) M. Hon and R. F. Davis: "Self-diffusion of ^{14}C in polycrystalline β-SiC," J. Mater. Sci., **14** (1979) 2411-2421.

62) M. Hon, R. F. Davis and D. E. Newbury : "Self-diffusion of ^{30}Si in polycrystalline β-SiC," J. Mater. Sci., **15** (1980) 2073-2080.

63) A. Seeger : "On the theory of the low temperature internal friction peak observed in metals," Phil. Mag., **1** (1956) 651-652.

64) P. L. Farnsworth and R. L. Coble : "Deformation behavior of dense polycrystalline SiC," J. Am. Ceram. Soc., **49** (1966) 264-268.

65) T. L. Francis and R. L. Coble : "Creep of polycrystalline silicon carbide," J. Am. Ceram. Soc., **51** (1968) 115-116.

66) R. N. Ghoshtagore and R. L. Coble : "Self-diffusion in silicon carbide," Phys. Rev., **143** (1966) 623-626.

67) D. P. Hasselman and H. D. Batha : "Strength of single crystal silicon carbide," Appl. Phys. Lett., **2** (1963) 111-113.

68) P. T. B. Shaffer and C. K. Jun : "The elastic modulus of dense polycrystalline silicon carbide," Mater. Res. Bull., **7** (1972) 63-69.

69) A. Djemel, J. Cadoz and J. Philibert : "Deformation of polycrystalline α-SiC," Proc. Conf. Creep and Fract. Eng. Materials and Structures, ed. by B. Wilshire and D. R. J. Owen, North Holland, Amsterdam (1981) 381-394.

70) C. Carry and A. Mocellin : "High temperature creep of dense fine grained silicon carbides," in Deformation of Ceramic Materials II, ed. by R. E. Tressler and R. C. Bradt, Plenum Publishing Corp., New York (1984) 391-404.

7. The Glassy State

James F. SHACKELFORD*

Abstract

The glassy state is defined in terms of atomic-scale structure, viz., short-range order (building blocks) and long-range randomness. On the microstructural-scale, these non-crystalline solids may exhibit phase separation. Optical properties (such as the transmission of visible, uv, and ir radiation) are central to many of the applications of glasses. At moderate temperatures, glasses are elastic structural materials susceptible to failure due to the action of tensile stresses on surface defects. At elevated temperatures (above the "glass transition"), glasses deform by viscous flow. Inert gas transport is a useful structural probe. Ionic transport is the basis of relatively high chemical durability and low electrical conductivity.

Keywords: glass, non-crystalline, structure, properties.

7.1 Introduction

Glass is a material both ancient and contemporary. Along with pottery, glassware was one of man's earliest durable products and provides a link to ancient civilizations. Contemporary glasses play a central role in some of our most sophisticated technologies. Glasses can be defined as "non-crystalline solids with compositions comparable to the crystalline ceramics."[1] In reference to fine ceramics, we can consider three aspects of the glassy state. First would be a glassy phase in a predominantly crystalline ceramic. Second would be a completely non-crystalline material, i.e., a true glass. And third would be a glass-ceramic in which the glassy state is primarily a precursor to a final, fully-crystallized product.

A significant glassy phase at the grain boundaries of a crystalline ceramic is largely associated with traditional ceramic technology, e.g., relatively large impurity levels and/or a significant amount of silica in the material composition.[2] More sophisticated fine ceramics are structurally more similar to common metals, in which crystallinity of the grains extends completely up to the grain boundary.[3] The structure of such interfacial regions can be modelled with tools developed for characterizing non-crystalline solids, but the regions do not represent a bulk glassy phase.[4] The presence of some glassy phase in certain ceramics is, then, acknowledged, but the "glassy state," for the sake of this review, will be discussed in terms of glass as a separate material. Similarly, this review will not treat further the

*Division of Materials Science and Engineering, University of California, Davis, California 95616, U.S.A.

subject of glass-ceramics. The interested reader can find that material discussed in some detail by McMillan.[5]

Traditional glasses involve a silica-based chemistry. Non-silicate oxide glasses are generally of little commercial value. However, there is currently substantial interest in various nonoxide glasses. Rapid solidification techniques developed for amorphous metals allow virtually any material composition to be formed in a non-crystalline or glassy state. Nonetheless, the primary examples to be used in this review for illustration of the nature of the glassy state will be vitreous silica and related silicate glasses. These serve as classic examples of glasses and cover the majority of the commercial glass industry, including some of the most sophisticated technological applications.

A single review paper cannot, of course, adequately cover the range of current understanding about an important material such as glass. Various monographs are available for more detailed study. The field of "glass science" was systematically reviewed by R.H. Doremus in 1973.[6] Two volumes in the monograph series entitled "Treatise on Materials Science and Technology" have been subtitled "Glass I"[7] and "Glass II"[8] and contain several review articles edited by M.Tomozawa and R.H.Doremus. An ambitious series of monographs entitled "Glass: Science and Technology" has been undertaken by editors D.R.Uhlmann and N.J.Kreidl. The first of twelve planned volumes has been published covering the area of "Elasticity and Strength in Glasses."[9] Finally, an extensive collection of data in glass systems from the Russian literature is being translated into English. The first volume has been published under editor O.V.Mazurin.[10]

7.2 Structure

In Section 7.1, the basic definition of glass centered on its lack of crystalline order. The primary statement of the nature of this non-crystallinity was given in the classic paper

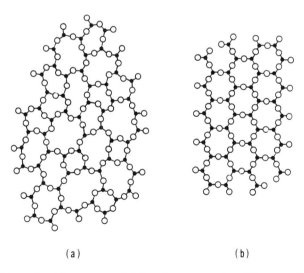

(a)   (b)

Fig. 7.1 Schematic of the structure of (a) a glass compared to (b) a crystal of the same composition. The two structures have the same building block (short-range order), but the glass lacks long-range order. [after W. H. Zachariasen[11]]

of Zachariasen.[11] The famous Zachariasen schematic is shown in Fig. 7. 1(a) in contrast to the corresponding crystalline structure in Fig. 7. 1(b). The key features of the non-crystalline schematic are (1) AO_3 triangular "building blocks" comparable to those in the crystalline material and (2) an irregular connectivity of the building blocks in contrast to the regular connectivity in the crystalline material. The building block is a component of short-range order (sro) in a material lacking in long-range order (lro). Zachariasen argued that the connectivity of building blocks is essentially random in nature. This "long-range randomness" (lrr) can be specified in terms of distributions of structural parameters such as the A–O–A bond angle, the n-membered oxygen ring, and interstitial size. The significance of Zachariasen's achievement in providing a useful working definition of glass structure, which is still widely used over 50 years later, is especially impressive in that it was solely an intellectual exercise based on his understanding of the principles of crystal chemistry. Zachariasen's "random network theory" became widely accepted following the publication of supporting experimental evidence by Warren and co-workers in 1936.[12] Their early use of the Fourier transform of X-ray scattering data confirmed the presence of SiO_4 tetrahedra in vitreous silica with an apparently random linkage of those tetrahedra. Figure 7. 2

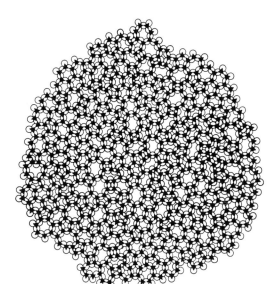

Fig. 7. 2   A 300-ring "triangle raft" is an extended Zachariasen schematic that allows statistical analysis of structural parameters. [after J. F. Shackelford[13]]

represents an extended Zachariasen schematic[13] with enough interstices (oxygen rings in this two-dimensional schematic) to allow a statistical analysis of the distribution of their size (Fig. 7. 3). The histogram of interstitial size (Fig. 7. 3(a)) shows a large, single bar for the 3-coordinated interstices associated with the AO_3 building block and a distribution of sizes for the randomly-generated rings. A best-fit of the continuous distribution of interstitial size (Fig. 7. 3(b)) becomes a sharp spike representing the 3-coordinated A^{3+} ion (short-range order) and a lognormal distribution representing the long-range randomness. This overall distribution (Fig. 7. 3(b)) is, then, a working definition of an ideally random

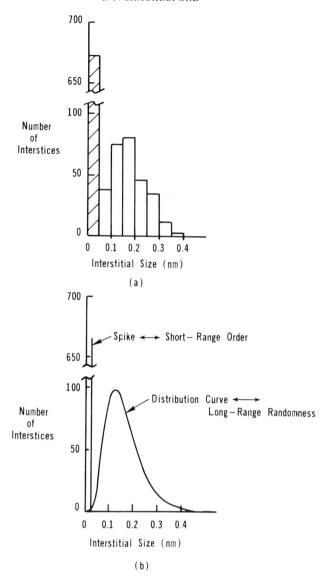

Fig. 7.3 (a) Histogram of interstitial size (defined as an inscribed-circle) for the triangle raft of Fig. 7.2. There is large bar associated with the 3-membered rings of the AO_3 building blocks. (b) A best-fit of the continuous distribution of interstitial size serves as a working definition of an ideally random structure. [after J. F. Shackelford[13]]

structure. The random network theory has been a widely used and generally adequate model of the structure of vitreous silica. There has been, however, an ongoing debate about the ideality of the randomness. Substantial discussion has been given to experimental evidence for ordering in this non-crystalline material.[14] Various spectroscopic techniques have been effective in providing this more complete structural picture. Although the validity of the random network model is the source of substantial debate and current research for vitreous silica, structural studies on more chemically-complex silicate glasses during the past two decades have shown the random network model to be inadequate for such materials. As a simple example, a random network model of an alkali modified silicate (such as $x\mathrm{Na_2O} \cdot (1-x)\mathrm{SiO_2}$) would involve a random distribution of "modifier" $\mathrm{Na^+}$ ions within the silica network (shown schematically in Fig. 7. 4[15]). However, X-ray diffraction studies of such glasses (using the techniques pioneered by Warren[12]) indicate that the average $\mathrm{Na^+}$-$\mathrm{Na^+}$ separation distance is substantially less than would be the case for random distribution.[16]

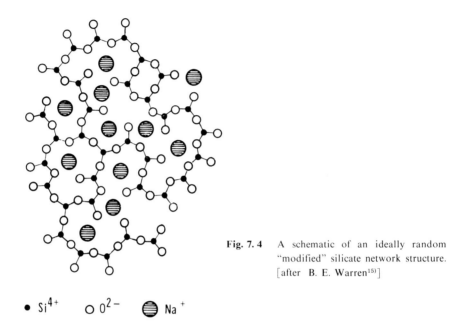

Fig. 7. 4 A schematic of an ideally random "modified" silicate network structure. [after B. E. Warren[15]]

• $\mathrm{Si^{4+}}$ ○ $\mathrm{O^{2-}}$ ⊜ $\mathrm{Na^+}$

Especially dramatic evidence of ordering has been demonstrated in vitreous $\mathrm{B_2O_3}$ and alkali-modified borate glasses. Although these materials are of minor commercial significance, they are attractive for model structural studies. In Warren's pioneering study of vitreous silica,[12] he also identified $\mathrm{BO_3}$ triangular building blocks in vitreous $\mathrm{B_2O_3}$. This implied that the $\mathrm{B_2O_3}$ structure would be directly analogous to the Zachariasen schematic of Fig. 7. 1(a), although not constrained to two dimensions, of course. However, nuclear magnetic resonance (NMR) studies by Krogh-Moe[17,18] indicated that a substantial concentration of planar $\mathrm{B_3O_6}$ boroxyl rings are present in the network structure (see Fig. 7. 5). Substantial refinement of the X-ray diffraction experiment by Mozzi and Warren[19] confirmed this structural model. Furthermore, NMR studies of various alkali-modified borate glasses have shown that modest alkali levels lead to the formation of some 4-coordinated boron ions in $\mathrm{BO_4}$ tetrahedra.[20] The fraction of boron ions in 4-fold coordination reaches a maximum of 0.5 at approximately 40 mole percent alkali oxide addition. Further modifier addition leads to a steady decrease in the fraction of 4-fold coordination (see Fig. 7. 6).[20] The complexity of the structural ordering in these relatively simple model glasses is an indication of the inadequacy of the random network model for general, commercial glass compositions.

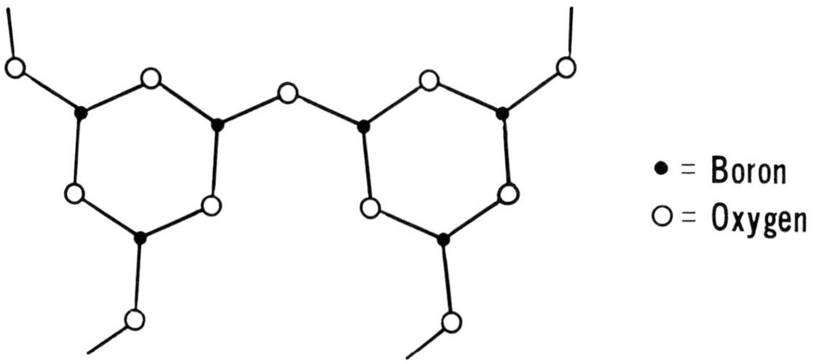

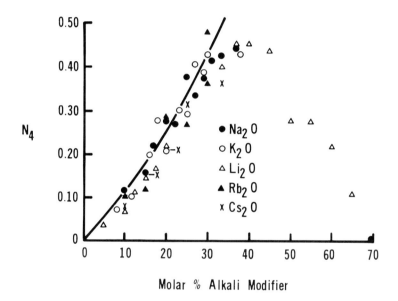

Fig. 7.5 A schematic model of boroxyl rings in B_2O_3. [after R. L. Mozzi and B. E. Warren[19]]

Fig. 7.6 The fraction (N_4) of boron ions in four-fold coordination (BO_4 tetrahedra) in alkali borate glasses plotted as a function of the molar percent of alkali oxide. [after P. J. Bray[20]]

Our discussion of glass structure has centered on atomic-scale geometry consistent with the definition of glass in terms of its non-crystallinity. It is also important to note that, on the microstructural-scale, many glasses show a tendency to phase separate.[21] Fig. 7.7 illustrates a typical phase-separated microstructure. This phenomenon, analogous to liquid-liquid immiscibility, can be modelled in terms of both thermodynamic principles and structural considerations derived from crystal chemistry. The practical significance of this phenomenon is the effect of the resulting microstructure on certain properties, especially within the optical and transport categories.

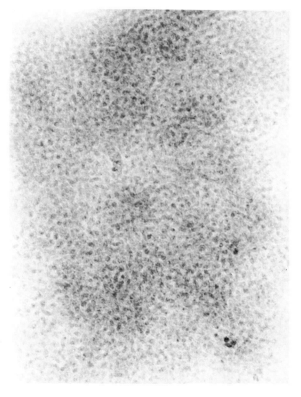

Fig. 7.7 Transmission electron micrograph of phase separation in a 20 mole percent Al_2O_3 - 80 mole percent SiO_2 glass, 100,000 X. (Courtesy of P. J. Hood)

7.3 Properties

With an understanding of the nature of the atomic- and microscopic-scale structure of glasses, we can proceed with a systematic discussion of their key properties. These are divided into three broad categories: optical, mechanical, and transport. Taken together, these properties determine the bases of the application of these glassy materials in modern technology and also the limitation of such applications. It is important to note that our knowledge of structure gained in Section 7.2 is essential in understanding these properties. Conversely, we shall find that property measurements can be an important tool in improving our understanding of glass structure.

7.3.1 Optical

Perhaps the most distinctive feature of traditional glassware is its optical transparency. This and related optical properties are at the foundation of many of the contemporary applications of glasses, such as windows, lenses, containers, filters, lasers, and waveguides.[22] Various applications are dependent on the percentage transmission as a function of wavelength within (1) the visible region (0.4-0.7 μm wavelength), (2) the ultraviolet (<0.4 μm), and (3) the infrared (>0.7 μm). Figure 7.8 shows a typical transmission curve in the infrared region for vitreous silica, which is, of course, transparent in the visible range but

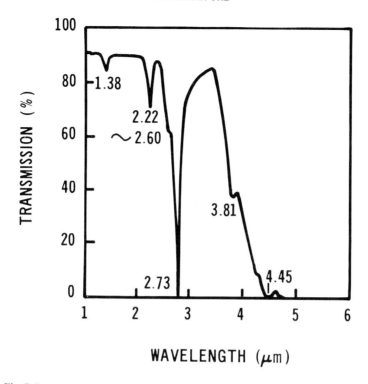

Fig. 7.8 Typical transmission curve for vitreous silica in the infrared region.

drops sharply in transmission for wavelengths between 4 and 5 μm. The various absorption peaks between 1 and 3 μm are associated with chemically dissolved water (in the form of unassociated OH groups). Especially distinctive is the 2.73 μm peak associated with a fundamental vibrational mode. The magnitude of this peak is the basis for determining the "water" content of this glass. The coloration of glasses results from similar, selective absorption bands within the visible range. The specific mechanism for such absorption is the excitation of 3d electrons in partially filled inner shells of transition metal elements (Ti, V, Cr, Mn, Fe, Co, Ni, and Cu). Table 7.1 summarizes the colors associated with various transition metal ions.

Phase separation can be the source of light scattering due to different indices of refraction associated with each phase.[21] A quantitative analysis of scattering can be an effective tool for identifying the immiscibility region in binary or multi-component glasses.[23]

The unique optical behavior of photochromic glass is the source of commercial application in the ophthalmic market (i.e., for "eyeglasses").[24] These materials change color (usually darkening) when exposed to light. The color change is reversible. A typical mechanism in a homogeneous glass involves sufficient energy from an ultraviolet photon to produce an electronic defect and resulting light absorption. Reversibility comes from electron mobility restoring the original electronic structure. There is a technological advantage to the use of nonhomogeneous glasses with a fine suspension of crystallites, viz., the glass properties can be adjusted independently of the photochromism (determined by

THE GLASSY STATE

Table 7.1 Coloration due to transition metal ions in glasses. [after G. H. Sigel, Jr.[22]]

Configuration	Ion	Color
d^0	Ti^{4+}	colorless
	V^{5+}	faint yellow to colorless
	Cr^{6+}	faint yellow to colorless
d^1	Ti^{3+}	violet-purple
	V^{4+}	blue
	Mn^{6+}	colorless
d^2	V^{3+}	yellow-green
d^3	Cr^{3+}	green
d^4	Cr^{2+}	faint blue
	Mn^{3+}	purple
d^5	Mn^{2+}	light yellow
	Fe^{3+}	faint yellow
d^6	Fe^{2+}	blue-green
	Co^{3+}	faint yellow
d^7	Co^{2+}	blue-pink
d^8	Ni^{2+}	brown-purple
d^9	Cu^{2+}	blue-green
d^{10}	Cu^+	colorless

the crystallites). Figure 7.9 illustrates the spectral response of a slab of a glass with copper-cadmium halide precipitates. The darkening mechanism is believed to be associated with a solution of cuprous ions in a cadmium chloride crystallite. Photolysis is accompanied by the production of cuperic ions in nonequilibrium sites giving the glass a

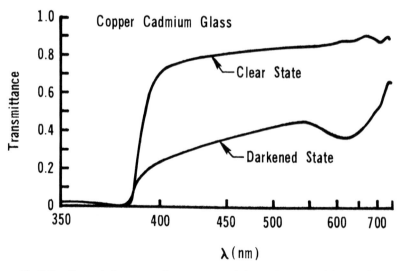

Fig. 7.9 Transmission curves for a copper-cadmium (photochromic) glass in clear and darkened states. [after R. J. Araujo[24]]

green color. Among the most widely used photochromic glasses are those using silver halide precipitates. The darkening and bleaching mechanisms of these photochromic glasses are believed to be quite analogous to the photolytic effects found in bulk silver halide crystals. In addition to ophthalmic applications, photochromic glasses can be expected to find uses in architecture and information storage and display.

Although the photochromic effect is a useful application, radiation effects on other glasses can frequently be detrimental, especially in relation to optimal transmission requirements. Friebele and Griscom[25] have thoroughly reviewed the literature in this area. They have systematically cataloged the various types of local structural defects produced by radiation in a wide range of glass compositions.

7.3.2 Mechanical

The wide use of glass as a structural component makes the mechanical properties of fundamental importance. Recent years have seen rapid progress in understanding and controlling these properties.[26] To provide a comprehensive view of the mechanical behavior of glass, it is necessary to consider a wide temperature range. Glasses are characteristically brittle at moderate temperatures but readily deformable at elevated temperatures.

Vitreous silica near room temperature is a nearly ideal elastic material,[27] i.e., it does not exhibit creep deformation and recovers instantly after prolonged deformation. However, the addition of modifier oxides such as Na_2O and CaO can lead to both creep and delayed recovery. Ernsberger[27] distinguishes such deformation from true viscous flow which occurs at higher temperatures (with attendant viscosities greater than 10^{13} poise). For those glasses that tend to phase separate, elastic properties can be modelled with traditional analytical techniques associated with composite systems.

Substantial practical information about the mechanical behavior of structural glasses has been obtained within the past decade by the application of fracture mechanics,[28] which can be defined as the introduction of a large crack into the material followed by analysis of material resistance to propagation of the crack under various environments and stresses. Knowledge of the nature of crack propagation and flaw geometry provides for the prediction of strength and/or lifetime of given structures. The most general fracture mechanics parameter is the "fracture toughness," or K_{IC}, which is the critical value of the stress intensity factor for a slit crack in an infinitely wide plate loaded under a tensile stress (perpendicular to the crack face). In general, the value of fracture toughness is approximated by

$$K_{IC} \cong \sigma_f \sqrt{\pi a} \qquad (7.1)$$

where σ_f is the applied tensile stress at failure and a is the length of a surface crack. Table 7.2 compares typical values of K_{IC} for various structural materials including glass.[29] Note that, in general, relatively "brittle" ceramics and glasses have relatively low values of K_{IC}. The predominant source of failure in glass is surface flaws produced by machining or handling. The distinctive fracture surface emerging from the initiating crack permits detailed "failure analysis" of a structural piece. A precise mechanism of fracture can generally be assigned using microscopic inspection. This technology has been reviewed by

Table 7.2 Typical values of fracture toughness (K_{1C}) for various materials including glass. [after M. F. Ashby and D. R. H. Jones[29]]

Material	K_{1C} (MPa$\sqrt{m}$)
Mild steel	140
Medium-carbon steel	51
Cast iron	6–20
Pure ductile metals (e.g., Cu, Ni, Ag, Al)	100–350
Be (brittle, hcp metal)	4
Aluminum alloys (high strength-low strength)	23–45
Titanium alloys (Ti 6 Al 4V)	55–115
Polyethylene	
High-density	2
Low-density	1
Polystyrene	2
Polypropylene	3
Electrical porcelain	1
Alumina (Al_2O_3)	3–5
Magnesia (MgO)	3
Cement/concrete, unreinforced	0.2
Silicon carbide (SiC)	3
Silicon nitride (Si_3N_4)	4–5
Soda glass (Na_2O-SiO_2)	0.7–0.8

Frechette.[30] Of special importance in the prediction of structure lifetimes is consideration of delayed failure, or "static fatigue." Fracture mechanics has helped to identify this phenomenon with a mechanism of subcritical growth of cracks, usually in the presence of water.[28]

The susceptibility of glass to failure by propagation of surface cracks under tensile loading has led to technologies of "strengthening" by means of either eliminating surface defects or by producing residual compressive stresses in the glass surface. Although these technologies have been used for several decades, advances in glass science in recent years have allowed significant improvements. Ernsberger[31] has reviewed these various technologies. Gardon[32] has discussed in greater detail the specific technique of "tempering" in which the residual surface compressive state is the result of careful heat treatment. Similarly, Bartholomew and Garfinkel[33] have discussed "chemical strengthening" in which chemical exchange at the glass surface (e.g., K^+ for Na^+) produces a compressive stress in the silicate network.

As noted earlier, glass exhibits viscous deformation at relatively high temperatures. This can be conveniently illustrated with a typical thermal expansion curve for glass shown in Fig. 7.10. The thermal expansion coefficient (slope of the $\Delta L/L$ versus T plot) below T_g is comparable to that of a crystalline solid of the same composition. Below T_g, the material is a true glass (a rigid solid), and T_g is referred to as the "glass transition

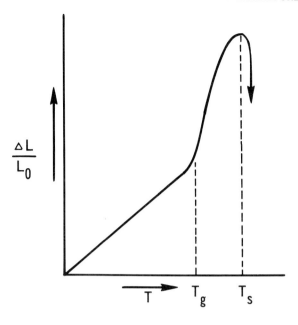

Fig. 7.10 A typical thermal expansion curve for a glass. Above T_g (the "glass transition temperature"), the material deforms viscously. [after J. F. Shackelford[1]]

temperature." Above T_g, the material is a supercooled liquid and deforms viscously. The precipitous drop in data at T_s (the "softening temperature") indicates that the material is so fluid that it can no longer support the weight of the length monitoring probe (a small refractory rod). The viscosity of a typical sode-lime-silica glass from room temperature to 1,500 °C is summarized in Fig. 7.11. A good deal of useful processing information is contained in such a plot relative to the manufacture of glass products. The annealing point

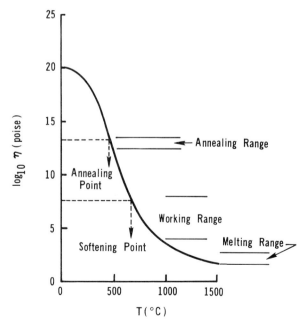

Fig. 7.11 Viscosity of a typical soda-lime-silica glass from room temperature to 1,500°C. [after J. F. Shackelford[1]]

(at which the viscosity is $10^{13.4}$ poise) corresponds approximately to the glass transition temperature. Above T_g, the viscosity data follow an Arrhenius form with

$$\eta = \eta_0 e^{+Q/RT} \tag{7.2}$$

where η_0 is the preexponential constant, Q is the activation energy for viscous deformation, and RT has the usual meaning. Viscosity deviates from this Arrhenius form below T_g as elastic deformation mechanisms come into play, and a plateau of 10^{20} poise is reached near room temperature. Ernsberger's caution about describing such deformation strictly in terms of a viscosity parameter[27] should again be noted.

7.3.3 Transport

Frequently, the applications of glasses are dependent on certain transport properties. Alternately, applications may be limited by such transport properties.

A relatively simple transport system is the diffusion and solution of unreactive gases in glass. Studies of these systems have been reviewed by Shelby[34] and Shackelford.[35,36] The relatively open structure of silica-based glasses permits measurable diffusion of numerous gases. Substantial structural information is available from the careful analysis of this transport, as indicated schematically by Fig. 7.12. An example of the correlation of transport parameters with a phase separation boundary is illustrated by the Na_2O-SiO_2 system in Fig. 7.13. Rothman et al.[37] have shown that, for small concentration levels, modifier ion diffusion can be modelled as an interstitial mechanism through the silicate network structure (analogous to unreactive gas atom transport). The general case of ionic transport in high modifier concentration glass systems has been reviewed by Doremus[6] and Kingery et al.,[38] including the relationship of ionic diffusion to the electrical conductivity

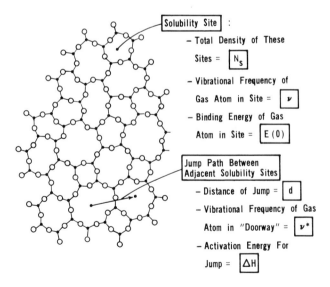

Fig. 7.12 Solubility and diffusion parameters for inert gas transport are related to the structure of the glass network. [after J. F. Shackelford[35]]

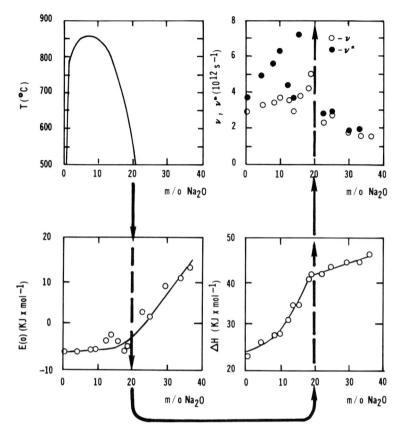

Fig. 7.13 The transport parameters defined in Fig. 7.12 correlate with phase separation in the Na_2O-SiO_2 system. [after J. F. Shackelford[35)]]

of these materials. The important relationship of ionic diffusion to the chemical durability of glass has also been reviewed by Doremus.[39)] He has discussed the reaction of typical alkali silicate glasses with water in terms of the interdiffusion of alkali and hydronium ions. The effect of glass composition on chemical durability can, in turn, be discussed in terms of changes in the diffusion coefficients for these ionic species. As with other properties, phase separation can have a strong influence. Ionic diffusion is, then, a transport property associated with the limitation of glasses in certain applications. On the other hand, this limitation is a relative one. The small magnitude of ionic diffusivities in glasses at moderate temperatures makes these materials among the more durable for common structural applications. The antique glassware mentioned at the outset of this review is dramatic testimony to this fact.

7.4 A Future Need: Structural Classification

To this point, our review has emphasized structure-property relationships for glass. Next, it is appropriate to look forward to areas of current research which can have the strongest impact on future applications of this material. A technological area of future significance is the fabrication of glass by the sol-gel method.[40)] By avoiding conventional starting materials, glass can be fabricated at substantially lower temperatures. In addition, novel glass compositions are possible. Whatever production methods are used for future glass products, a pressing need remains to better understand the nature of the atomic-scale structure of glass. Much of the controversy dealing with glass structure (the most active area of glass research in the past two decades) stems from the lack of generally accepted "ground rules" for the nature of that structure.

In the more than 50 years since Zachariasen's classic paper on glass structure,[11] a massive quantity of information has been acquired about the structure of non-crystalline solids. However, there has been surprisingly little progress in the direction of defining a structural classification system analogous to crystallography. Only as recently as 1980 was the term "amorphography" presented by Wright et al.[41] as an appropriate title for "the branch of physical science dealing with the structures of amorphous materials and their systematic classification." In that paper, the utility of structural transformations was emphasized to illustrate the correspondence between the apparently distinct random network model of oxides and the random close packed model of amorphous metals. The purpose of this section is to further explore the structural characteristics necessary for such a systematic classification. The specific approach will be twofold: (i) to propose a working definition of an ideally random structure analogous to the ideally ordered structures (space lattices) of crystallography and (ii) to identify the set of building blocks associated with short range order.

Zachariasen stated that in 1932 "... we know practically nothing about the atomic arrangement in glasses." More than fifty years later, we can, at least, acknowledge two fundamental facts about non-crystalline solids: (i) all lack significant long-range order and (ii) some have detectable short-range order. The two-pronged classification system of this section attempts to quantify both observations.

Substantial work has been done in the last two decades on modeling amorphous metals. Among the more sophisticated studies was that by Finney and Wallace[42]. An interesting example of a computer-generated interstitial distribution from that paper is given in Fig. 7. 14. This distribution of inscribed spheres bears a strong resemblance to the skewed form of the lognormal distribution shown in Fig. 7. 3 (b). Figure 7. 14 summarizes a reasonably close approximation of the lognormal distribution to the computer-generated histogram.

This result is reinforcement for the hypothesis of Section 7. 2 that the lognormal distribution of interstitial size is a working definiton of ideal structure. Kurtz[43] has suggested that the log-normal distribution function is a likely candidate for a general description of the random subdivision of space.

The histogram of Fig. 7. 14 was generated with a hard-sphere interatomic potential model. An interesting result of using a soft sphere (Lennard-Jones) potential is shown in Fig. 7. 15.[42] Finney and Wallace found that the resulting bimodal distribution corresponds to a concentration of tetrahedral (T) and octahedral (O) interstices. In effect, the "squashiness," as they termed it, provided by the soft potential causes a full range of canonical holes[44] (suggested by Fig. 7. 14) to give way to an assembly of only tetrahedra and octahedra, albeit somewhat distorted. Finney and Wallace further assert that such a

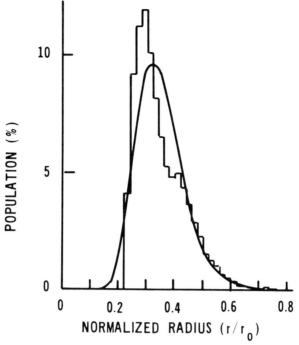

Fig. 7.14 Computer-generated distribution of interstitial size (as indicated by inscribed sphere radius divided by atomic radius) for a hard-sphere model of an amorphous metal[42] and the resulting best-fit of a lognormal distribution to the histogram.

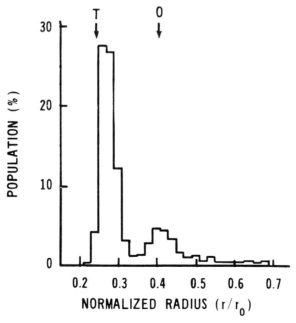

Fig. 7.15 A "soft sphere" model of an amorphous metal leads to a bimodal distribution of interstitial size,[42] an indication of structural ordering.

bimodal structure is "ideal," although they acknowledge that the hard-sphere (random close packed) structure is a more traditional model of ideality for amorphous metals. In the perspective of the current paper, I must endorse the random close packed structure (which gives the interstitial distribution of Fig. 7. 14) as the ideal structure. Finney and Wallace prefer the simplicity of a structure composed of only two types of interstices. However, this simplicity does not necessarily imply ideality. One should recall that crystalline close-packed structures are composed entirely of tetrahedral and octahedral holes. The perspective of the current paper would define the bimodal distribution of Fig. 7. 15 as an example of a "defect" structure (ordering) relative to an essentially ideal (random) structure given in Fig. 7. 14. Such a perspective is consistent with experience in modeling grain boundary structures[44] in which the arrangement of canonical holes is more ordered for soft sphere potentials than for hard sphere potentials.

Perhaps the more significant implication of the soft-sphere potential results (Fig. 7. 15) is that real amorphous metals (which will, of course, have "soft" rather than "hard" potentials) may have a significant tendency toward ordering.*1 An advantage of a specific definition of ideally random structure, such as that provided by Section 7. 2, is the ability to quantify the degree of ordering.

Zachariasen's rules of glass structure include the observation that building blocks in common glass-forming systems are limited to small oxygen polyhedra. More specifically, Zachariasen restricted his attention to oxygen "triangles" and "tetrahedra," which were subsequently identified by Warren et al.[12] in B_2O_3 and SiO_2 glasses, respectively.*2 There are four types of building block geometry for non-crystalline solids. By definition, the structural "units" must be either zero-dimensional (type 0), one-dimensional (type 1), two-dimensional (type 2), or three-dimensional (type 3). Zachariasen's triangle and tetrahedron would be examples of types 2 and 3, respectively. The monatomic (type 0) unit is the basis of amorphous metals. Zachariasen did not consider the monatomic system, of course, because of the lack of evidence for amorphous metals in the 1930's. The one-dimensional (type 1) unit would not be expected to form a network structure, as implied by Zachariasen's rule 4,[11] but is the basis of non-crystalline linear polymers. Polyethylene is a common example.

Although any structural building block must be of one of the four types summarized above, an additional factor is necessary for full characterization of short-range order. Specifically, one must also identify the dimensionality of the building block connectivity. Newnham et al.[45] have used this concept in characterizing the geometry of composite materials. As with building block geometry, connectivity can be zero-dimensional (or non-directional = type 0), one-dimensional (type 1), two-dimensional (type 2), or three-dimensional (type 3). The amorphous metal would have both type 0 geometry and type 0 connectivity. The linear polymer polyethylene would have type 1 geometry and con-

---

*1 This implication is similar to the well-documented[19] evidence for nonrandomness in the ring structure of B_2O_3 glass, viz., the common occurrence of boroxyl rings rather than a random network.

*2 Building blocks with connectivity larger than four were not considered by Zachariasen. Such AO_n ($n > 4$) polyhedra could form non-crystalline solids theoretically, perhaps by rapid quench techniques. They are not, however, associated with common glass-forming systems, as noted in Zachariasen's rule 2.

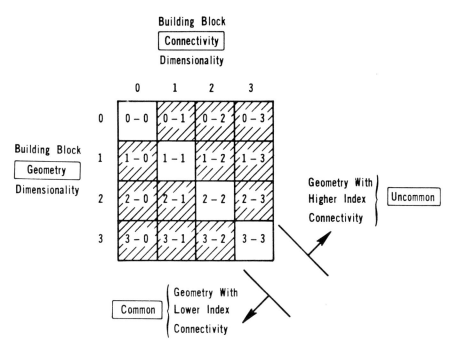

Fig. 7.16 A simple matrix indicating that any non-crystalline solid can be categorized into one of 16 types based on the dimensionality of both the geometry and connectivity. Some of the most common non-crystalline solids fall along the diagonal, e. g., amorphous metals (type 0-0) and vitreous silica (type 3-3). Non-crystalline solids also commonly occur with connectivity having a lower index than geometry, e. g., a chain silicate (type 3-1) involving a three-dimensional building block with only linear connectivity.

nectivity. B_2O_3 glass would have type 2 geometry and connectivity. Vitreous SiO_2 would have type 3 geometry and connectivity. However, geometry and connectivity will not always have the same index. Any non-cryctalline solid can be categorized in terms of its geometry/connectivity indices as summarized in Fig. 7.16. The solids along the diagonal of the matrix in Fig. 7.16 have equivalent geometry and connectivity indices. Thus, an amorphous metal is a type 0-0 solid and vitreous silica is type 3-3. A useful feature of Fig. 7.16, in addition to providing a compact notation system, is to indicate that those m-n glasses below the diagonal (i. e., $m > n$) are relatively common. A sheet silicate would be a 3-2 type, a chain silicate would be a 3-1 type, and so on. Conversely, those m-n glasses above the diagonal ($m < n$) are relatively uncommon. In those cases, the connectivity is of a higher index than the geometry. Such bonding configurations are physically possible but not likely. In any case, Fig. 7.16 provides a compact set of 16 possible short-range order configurations for all non-crystalline solids analogous to the comprehensive set of 14 space lattices of crystallography.

7.5 Conclusions

The glassy state is defined in terms of a lack of crystalline structure. Some crystalline ceramics contain a significant glassy phase. The glassy state is also a precursor in the production of glass-ceramics. Structural studies during the past five decades have refined

Zachariasen's original model of glass consisting of short-range order (building blocks) and long-range randomness. Phase separation is a microstructural-scale phenomenon of considerable importance to the nature of many glass properties. Optical transmission is one of the central properties in both traditional and advanced applications of glasses. At moderate temperatures, glasses are characteristically elastic materials, and fracture mechanics has permitted improved strength and lifetime prediction. The susceptibility of glass to failure by the propagation of surface cracks under tensile stresses has led to various strengthening technologies. The surface defects can be minimized, or a residual compressive stress can be produced ("tempering" or "chemical strengthening"). At elevated temperatures (above the "glass transition"), glass deforms in a viscous manner, with the viscosity following an Arrhenius form. Various transport properties play a central role in the application of glasses. Inert gas transport can be a sensitive structural probe. Ionic transport is a dominant factor in the relatively high chemical durability and low electrical conductivity of glasses. A future need in glass research is a structural classification system for non-crystalline solids.

Acknowledgements

This work has been made possible with the support of National Science Foundation Grant DMR 82-04394.

References

1) J. F. Shackelford: Introduction to Materials Science for Engineers, MacMillan, New York (1985) Chap. 8.
2) A. L. Stuijts: "Ceramic microstructures," in Ceramic Microstructures '76, ed. by R. M. Fulrath and J. A. Pask, Westview Press, Boulder, Colorado (1977) 1-26.
3) R. W. Balluffi, P. D. Bristowe and C. P. Sun: "Structure of high-angle grain boundaries in metals and ceramic oxides," J. Am. Ceram. Soc., **64** (1981) 23-34.
4) J. F. Shackelford: "A canonical hole model of grain boundaries and interfaces in oxides," in Advances in Ceramics, Vol. 6, Character of Grain Boundaries, ed. by M. F. Yan and A. H. Heuer, American Ceramic Society, Columbus, Ohio (1983) 96-101.
5) P. W. McMillan: Glass-Ceramics, 2nd ed., Academic Press, New York (1979).
6) R. H. Doremus: Glass Science, John Wiley and Sons, New York (1973).
7) M. Tomozawa and R. H. Doremus (eds.): Glass I: Interaction with Electromagnetic Radiation, Academic Press, New York (1977).
8) M. Tomozawa and R. H. Doremus (eds.): Glass II, Academic Press, New York (1979).
9) D. R. Uhlmann and N. J. Kreidl, (eds.): Glass: Science and Technology, Vol. 5, Elasticity and Strength in Glass, Academic Press, New York (1980).
10) O. V. Mazurin, M. V. Streltsina and T. P. Shvaiko-Shvaikovskaya: Handbook of Glass Data, Part A: Silica Glass and Binary Silicate Glasses, Elsevier Science Publishers, New York (1983).
11) W. H. Zachariasen: "The atomic arrangement in glass," J. Am. Chem. Soc., **54** (1932) 3841-3851.
12) B. E. Warren, H. Krutter and O. Morningstar: "Fourier analysis of X-ray patterns of vitreous SiO_2 and B_2O_3," J. Am. Ceram. Soc., **19** (1936) 202-206.
13) J. F. Shackelford: "Triangle rafts—Extended Zachariasen schematics for structure modeling," J. Non-Cryst. Solids, **49** (1982) 19-28.
14) Proc. Intl. Symp. on Bonding and Structure in Non-Crystalline Solids, Wash., D. C., May 1983, to be published by Plenum Press, New

York.

15) B. E. Warren: "Summary of work on atomic arrangement in glass," J. Am. Ceram. Soc., **24** (1941) 256-261.

16) E. A. Porai-Koshits: "The possibilities and results of X-ray methods for investigation of glassy substances," in The Structure of Glass, Vol. 1, Consultants Bureau, New York (1958) 25-35.

S. M. Ohlberg and J. M. Parsons: "The distribution of sodium ions in soda-lime-silica glass," in Proc. Conf. on Physics of Non-Cryst. Solids, ed. by J. A. Prins, John Wiley and Sons, New York (1965) 31-40.

G. Carraro and M. Domenici (cited in Wright, A. C.): "The structure of amorphous solids by X-ray and neutron diffraction," in Advances in Structural Research by Diffraction Methods, ed. by W. Hoppe and R. Mason, Pergamon Press, New York (1974) 59.

17) J. Krogh-Moe: "Intrerpretation of the infrared spectra of boron oxide and alkali borate glasses," Phys. Chem. Glasses, **6** (1965) 46-54.

18) J. Krogh-Moe: "The structure of vitreous and liquid boron oxide," J. Non-Cryst. Solids, **1** (1969) 269-284.

19) R. L. Mozzi and B. E. Warren: "The structure of vitreous boron oxide," J. Appl. Cryst., **3** (1970) 251-257.

20) P. J. Bray: in Magnetic Resonance, ed. by C. K. Cougan, N. S. Ham, S. N. Stewart, J. R. Pilbrow and G. V. H. Wilson, Plenum Publishing Corp., New York (1970) 11-39.

21) M. Tomozawa: "Phase separation in glass," in Glass II, ed. by M. Tomozawa and R. H. Doremus, Academic Press, New York (1979) 71-113.

22) G. H. Sigel, Jr.: "Optical absorption of glasses," in Glass I, ed. by M. Tomozawa and R. H. Doremus, Academic Press, New York (1977) 5-89.

23) J. Schroeder: "Light scattering of glass," in Glass I, ed. by M. Tomozawa and R. H. Doremus, Academic Press, New York (1977) 157-222.

24) R. J. Araujo: "Photochromic glass" in Glass I, ed. by M. Tomozawa and R. H. Doremus, Academic Press, New York (1977) 91-122.

25) E. J. Friebele and D. L. Griscom: "Radiation effects in glass", in Glass II, ed. by M. Tomozawa and R. H. Doremus, Academic Press, New York (1979) 257-351.

26) D. R. Uhlmann and N. J. Kreidl: in the Preface to Glass: Science and Technology, Vol. 5, ed. by D. R. Uhlmann and N. J. Kreidl, Academic Press, New York (1980) page x.

27) F. M. Ernsberger: "Elastic properties of glasses," in Glass: Science and Technology, Vol. 5, ed. by D. R. Uhlmann and N. J. Kreidl, Academic Press, New York (1980) 1-19.

28) S. W. Freiman: "Fracture mechanics of glass," in Glass: Science and Technology, Vol. 5, ed. by D. R. Uhlmann and N. J. Kreidl, Academic Press, New York (1980) 21-78.

29) M. F. Ashby and D. R. H. Jones: Engineering Materials—An Introduction to Their Properties and Applications, Pergamon Press, Elmsford, New York (1980).

30) V. D. Frechette: "The fractography of glass," in Introduction to Glass Science, ed. by L. D. Pye, H. J. Stevens and W. C. La Course, Plenum Publishing Corp., New York (1972) 433-450.

31) F. M. Ernsberger: "Techniques of strengthening glasses," in Glass: Science and Technology, Vol. 5, ed. by D. R. Uhlmann and N. J. Kreidl, Academic Press, New York (1980) 133-144.

32) R. Gardon: "Thermal tempering of glass," in Glass: Science and Technology, Vol. 5, ed. by D. R. Uhlmann and N. J. Kreidl, Academic Press, New York (1980) 145-216.

33) R. F. Bartholomew and H. M. Garfinkel: "Chemical strengthening of glass," in Glass: Science and Technology, Vol. 5, ed. by D. R. Uhlmann and N. J. Kreidl, Academic Press, New York (1980) 217-270.

34) J. E. Shelby: "Molecular solubility and diffusion," in Glass II, ed. by M. Tomozawa and R. H. Doremus, Academic Press, New York (1979) 1-40.

35) J. F. Shackelford: "The potential of structural analysis from gas transport studies," J. Non-Cryst. Solids, **42** (1980) 165-174.

36) J. F. Shackelford: "Structural implications of gas transport in amorphous solids," in Proc. Int. Symp. on Bonding and Structure in

Non-Crystalline Solids, Wash., D. C. (May 1983), to be published by Plenum Press, New York.

37) S. J. Rothman, T. L. M. Marcuso, L. J. Nowicki, P. M. Baldo and A. W. McCormick: "Diffusions of alkali ions in vitreous silica," J. Am. Ceram. Soc., **65** (1982) 578-582.

38) W. D. Kingery, H. K. Bowen and D. R. Uhlmann: Introduction to Ceramics, 2nd ed., John Wiley and Sons, New York (1976).

39) R. H. Doremus: "Chemical durability of glass," in Glass II, ed. by M. Tomozawa and R. H. Doremus, Academic Press, New York (1979) 41-69.

40) J. D. Mackenzie: "Applications of sol-gel methods for glass and ceramics processing," in Ultrastructure Processing of Ceramics, Glasses, and Composites, ed. by L. L. Hench and D. R. Ulrich, John Wiley and Sons, New York (1984) 15-26.

41) A. C. Wright, G. A. N. Connell and J. W. Allen: "Amorphograpy and the modeling of amorphous solid structures," J. Non-Cryst. Solids, **42** (1980) 69-86.

42) J. L. Finney and J. Wallace: "Interstice correlation functions; a new sensitive characterization of non-crystalline packed structures," J. Non-Cryst. Solids, **43** (1981) 165-187.

43) S. K. Kurtz: private communication.

44) M. F. Ashby, F. Spaepen and S. Williams: "The structure of grain boundaries described as a packing of polyhedra," Acta Metall., **26** (1978) 1647-1663.

45) R. E. Newnham, D. P. Skinner and L. E. Cross: "Connectivity and piezoelectric-pyroelctric composites," Mater. Res. Bull., **13** (1987) 525-536.

8. Electrical Conductivity of Ceramic Materials

Donald M. SMYTH*

Abstract

The effect of chemical composition on the electronic and ionic conductivities of ceramic materials is reviewed. The properties of n-type and p-type semiconducting oxides are compared in terms of their defect chemistry with particular emphasis on the influence of non-stoichiometry and impurity additions. Ionically conducting solids are divided into intrinsic, extrinsic, and superionic conductors, and their properties are discussed in terms of both composition and structure.

Keywords : conductivity, ceramic, defects, ionic conductivity, semiconducting oxides.

8. 1 Introduction

The electrical conductivities of different materials cover one of the widest ranges of any observable physical property, from the essentially infinite conductivity of the superconductors to the infinitisimal conductivities of the best insulators. Ceramic materials are represented across the entire width of this spectrum. Some ceramics have electronic conductivities that rival those of metals, while others have ionic conductivities equivalent to those of the most highly conducting aqueous solutions. On the other hand, many ceramic materials find applications as capacitor dielectrics, circuit substrates, or device packages because of their extremely low conductivities. This review will discuss structural and compositional factors that are influential in determining the conductivity levels of ceramic materials.

8. 2 Charge Carrier Concentrations

The charge carrier concentration sets the general level of conductivity. Carrier mobilities may vary by several orders of magnitude, but even a high mobility cannot make up for an inadequate carrier concentration. Thus the energetic cost of producing a charge carrier is an important parameter. In intrinsic conductors this will be the enthalpy of intrinsic disorder. For electronic conductors the disorder reaction is the internal ionization reaction

$$\text{nil} \rightleftharpoons e' + h^{\cdot} \tag{8.1}$$

---

* Materials Research Center, No.32, Lehigh University, Bethlehem, Pennsylvania 18015, U.S.A.

where *nil* denotes the perfect crystal with all electrons in the lowest available energy states. e' and h˙ represent electrons and holes in conducting states, e.g. the conduction and valence bands. The enthalpy of Eq. (8.1) is frequently referred to as the band gap at 0 K, $E_g°$, and the carrier concentration is proportional to $\exp(-E_g°/2kT)$. The ionic analog of this thermally activated process would correspond to intrinsic Frenkel disorder in the silver halides

$$Ag_{Ag} + V_I \rightleftharpoons Ag_I^{\cdot} + V'_{Ag} \tag{8.2}$$

In this Kröger-Vink notation,[1] the main symbol represents the species, the subscript its location, and the superscript its charge relative to the ideal lattice, ˙and ' being extra positive and negative charges, respectively. V and I denote vacant and interstitial sites. The left-hand-side of Eq. (8.2) contains only normal lattice species. As in the electronic case, both defect species can contribute to conduction and the concentration of each of them is proportional to $\exp(-\Delta H_F/2kT)$, where ΔH_F is the enthalpy of Reaction (8.2).

The electron disorder reaction is, of course, the same for all solids, while the ionic disorder may be one of several types, e.g. cation Frenkel as described in Eq. (8.2), the analogous anion Frenkel, or Schottky disorder with vacancies in stoichiometric ratio on each sublattice. It is important to understand the energetic parameters that determine the extent of these types of disorder. Not only do these directly determine the concentrations of intrinsic defects, but they also give some indication of the extent to which extrinsic defects may result from impurities in solid solution.

The enthalpies of intrinsic disorder reactions always involve the formation of two defect species, or even more in the case of Schottky disorder in ternary or higher compounds. There is no indication of how much each defect species contributes to the total enthalpy; one may be energetically much more favorable than the other. Thus there can be a strong asymmetry in the extent of oxidation and reduction that can be achieved, or in the solubility of acceptor and donor impurities, since the different directions of chemical change will involve the formation of different defects. Consideration of the chemical properties of the elements present, and of the nature and crystal environment of possible ionic defects, can give insight into what is to be expected.

8.3 Electronic Conduction

8.3.1 General

Significant electronic conduction in a bulk solid oxide requires that the band gap be sufficiently small, generally less than 5 eV. This will occur in oxides or halides only when one of the cations has more than one achievable oxidation state, since the electronic states in the atomic species directly influence the energy states in the solid compound. Thus oxides of the main group elements such as the alkali metals, the alkaline earth metals, and the Group III elements B, Al, Sc, Y, and La are insulators. Only in the oxides of the transition metals including the lanthanides and actinides, and of the "inert pair" metals, are there opportunities for useful electronic conduction. (The "inert pair" refers to a chemical inertness of the two *s* electrons in such main group metals as Tl, Pb, Sn, and Bi that gives rise to two possible oxidation states, the group oxidation state and one that is two charges less, e.g., Pb^{+2} and Pb^{+4}, Tl^{+1} and Tl^{+3}.) In these modest band gap compounds there is an asymmetry in the conduction behavior depending on whether the cation is in its highest available oxidation state, such as in TiO_2 or Nb_2O_5, or in its lowest available oxidation state, such as in MnO or NiO. In some cases a type of amphoteric behavior is possible when both higher and lower oxidation states are possilbe, e.g. Fe^{+3} can be oxidized to Fe^{+4} or reduced to Fe^{+2}.

If the cation can only be reduced, e.g. Ti^{+4} or Nb^{+5}, corresponding to an addition of electrons, then extra electrons are easily tolerated in the solid oxide and the compound can become an n-type semiconductor. This can occur by chemical reduction, or by doping with a cation of higher oxidation state, a donor impurity. There is no corresponding stability for holes and when oxidized or acceptor-doped (an acceptor is a substitutional cation of lower charge) an insulating material is obtained. The opposite situation results when the cation can only be oxidized, corresponding to a loss of electrons, e.g. Ni^{+2} or Cr^{+3}. In this case holes are compatible with the electronic states and p-type semiconduction results from oxidation or acceptor-doping. Conversely, reduction or donor-doping results in an insulating material. If the cation can be both oxidized and reduced, semiconduction is to be expected in both directions of chemical change.

These chemical distinctions give a convenient basis for classification into three different classes, the n-type or reduction-type semiconductors, the p-type or oxidation-type semiconductors, and the intermediate, amphoteric materials that become good semiconductors by either oxidation or reduction. The defect chemistry that leads to the contrasting behavior of the first two classes will be compared in detail. This review will not deal with those compounds of transition metals in lower oxidation states that have metallic conduction because of partially filled bands.

8.3.2 N-Type Semiconductors

The n-type, or reduction-type, semiconductors are compounds containing a transition metal in its highest oxidation state, e.g. TiO_2, Nb_2O_5, WO_3, $BaTiO_3$, $LiNbO_3$, and Ca_2WO_4. These materials are normally insulators when equilibrated in air, because they then have a stoichiometric excess of oxygen that makes them p-type semiconductors at high temperatures, but insulators at room temperature. The oxygen excess is a result of the naturally occurring impurity content.[2] Since the cations in these compounds are in relatively high oxidation states, ≥ 4, the most likely impurities are predominantly of lesser charge, i.e. they are acceptor impurities such as Al^{+3}, Fe^{+2} or Fe^{+3}, Mg^{+2}, Ca^{+2}, etc. These are all very common elements and far exceed the natural abundances of the possible donor impurities such as Ta^{+5} or W^{+6}. An acceptor impurity oxide brings less oxygen per cation into the crystal and this usually results in a corresponding number of oxygen vacancies. Thus in the incorporation of Al_2O_3 into Nb_2O_5

$$Al_2O_3 \longrightarrow 2Al''_{Nb} + 3 O_0 + 2V_0^{\cdot\cdot} \qquad (8.3)$$

there is insufficient oxygen to fill the lattice. This makes possible the easy addition of a stoichiometric excess of oxygen by partial filling of the extrinsic (impurity-related) $V_0^{\cdot\cdot}$

$$V_O^{\cdot\cdot} + 1/2\ O_2 \rightleftharpoons O_O + 2\ h^{\cdot} \tag{8.4}$$

The oxygen atoms that enter the structure pick up electrons to become oxide ions, leaving holes in the valence band. These holes are free to conduct at high temperatures, $\gtrsim 600\ °C$. At lower temperatures, however, they are increasingly trapped by the acceptor impurity centers

$$Al''_{Nb} + 2h^{\cdot} \rightleftharpoons Al_{Nb} \tag{8.5}$$

because acceptor states are typically ~ 1 eV above the valence band when the constituent cation has no available higher oxidation states. This results in a p-type insulator at room temperature when these compounds have been processed in an oxidizing environment such as air. It should be reemphasized that this important property results specifically from the naturally-occurring impurity content.

On reduction, oxygen leaves the crystal as neutral atoms leaving behind vacant oxygen sites and free electrons

$$O_O \rightleftharpoons 1/2\ O_2 + V_O^{\cdot\cdot} + 2e' \tag{8.6}$$

In some cases, such as $LiNbO_3$, cation excess defects are preferred so that the lattice contracts to eliminate the vacant oxygen sites, forcing cations into sites they do not normally occupy.[3] Because of the availability of lower oxidation states for the cations in reduction-type semiconductors, the donor levels can be very close to the conduction band. In that case the electrons created by reduction may remain free to conduct down to room temperature or below. The lower oxidation state reflects a tolerance in the solid compound for free electrons ($Ti^{+4} + e^- = Ti^{+3}$). Thus the reduced alkaline earth titanates are black semiconductors.[2]

Figure 8.1 shows the conductivity of $BaTiO_3$ equilibrated with various oxygen partial pressures, P_{O_2}, at temperatures where the oxide can attain thermodynamic equilibrium with the surrounding gas phase. At each equilibration temperature, the conductivity minimum corresponds to the compositionally compensated, stoichiometric condition. The conductivity increases with higher values of P_{O_2} as the addition of a stoichiometric excess of oxygen increases the hole concentration according to Eq. (8.4). For pressures below that at the minimum the conductivity increases with decreasing P_{O_2} as oxygen loss increases the electron concentration according to Eq. (8.6). For samples cooled rapidly from P_{O_2} values in the latter range, the electron concentration differs little at 25 °C from that at the equilibrium temperature, and the mobility is actually a little higher.[4] For the oxygen excess compositions, however, the holes are essentially all trapped out on cooling, so that an insulating material results.

The P_{O_2} at the conductivity minimum thus divides the compound into two totally different properties. Two samples of $BaTiO_3$, for example, may have identical conductivities at 1,000 °C, but the conductivities may differ by a factor of 10^{10} at 25 °C because one is oxygen-deficient and n-type, while the other is oxygen excess and p-type. This enormous difference in conductivity can result from a compositional difference of only 5 ppm in the oxygen content.[5]

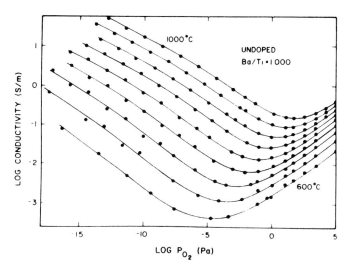

Fig. 8. 1 The equilibrium electrical conductivity of undoped polycrystalline $BaTiO_3$ with $Ba/Ti = 1,000$ as a function of oxygen partial pressure, P_{O_2}. The isothermal lines are at 50°C intervals from 600°C to 1,000°C. [Reproduced from Ref. 32) with permission from the authors and the American Ceramic Society.]

Semiconduction can often be achieved by donor-doping as well. This is because the donor impurity, having a higher charge than the ion it replaces, is trying to bring in more oxygen than the lattice can accommodate. If this extra oxygen is rejected to the gas phase, the loss of neutral oxygen atoms leaves behind the electrons with which they had been combined as oxide ions. In the case of $BaTiO_3$ doped with Nb_2O_5, the combination $BaO + (1-x) TiO_2 + xNbO_{2.5}$ ($= x/2$ Nb_2O_5) becomes $BaTi_{1-x}Nb_xe_x^-O_3 + x/4$ O_2 for $x < 0.005$. It is actually reduced by the loss of oxygen and is an n-type semiconductor. In defect notation the doping reaction can be written as

$$2BaO + Nb_2O_5 \longrightarrow 2Ba_{Ba} + 2Nb_{Ti}^{\cdot} + 6O_0 + 1/2\ O_2 + 2e' \qquad (8.7)$$

The extra positive charge of the donor impurity center is balanced by an equal number of extra electrons. For $x > 0.005$ in this system, the situation shown in Eq. (8. 7) is still valid after equilibration in a reducing atmosphere; however in an oxidizing atmosphere such as air, the extra oxygen is retained and the material is an insulator. It has recently been shown that in $BaTiO_3$ in the oxidized state this results from compensation of the donor impurity by titanium vacancies, V_{Ti}'''', rather than by electrons.[6,7] If necessary, a TiO_2-rich second phase will be formed to leave the proper concentration of V_{Ti}'''', and only the composition $BaTi_{1-5/4x}Nb_xO_3$ is single phase. When equilibrated with an oxidizing atmosphere the electronically-compensated, semiconducting state is limited to much smaller donor concentrations in $SrTiO_3$ and is apparently nonexistent in $CaTiO_3$.

The properties of the donor-doped material are thus dependent on whether the extra oxygen associated with the donor impurity is rejected to the gas phase, giving n-type semiconduction, or is retained in the lattice, giving an insulator. Oxgen loss is naturally favored by low P_{O_2} and high temperature. It was first shown for donor-doped $BaTiO_3$, $Ba_{1-x}La_xTiO_3$ and $BaTi_{1-x}Nb_xO_3$ with $0.005 < x < 0.020$, that the extra oxygen associated with the donor impurity can be reversibly gained or lost as a result of oxidation or reduction, respectively.[8] The weight change observed after successive equilibrations in air and CO at $1,060°C$ corresponded to the chemical equivalent of the charge on the impurity center, i.e. $x/2$ gram atoms of oxygen for the compositions described above. Typical experimental results are shown in Fig. 8. 2. This means that donor compensation corresponds to $[Nb_{Ti}^{\cdot}] = n$ in the reduced state, and to $[Nb_{Ti}^{\cdot}] = 4\,[V_{Ti}'''']$ in the oxidized state. The two states are connected by an oxidation-reduction equation

$$4xe' + xTi_{Ti} + xO_2 \underset{\text{reduction}}{\overset{\text{oxidation}}{\rightleftarrows}} xV_{Ti}'''' + xTiO_2 \qquad (8.8)$$

where the TiO_2 exsolved on oxidation will combine with $BaTiO_3$ to give $Ba_6Ti_{17}O_{40}$, the adjacent TiO_2-rich phase.[9-11] This demonstrates the general tendency for donor impurities to be compensated by electrons at low P_{O_2}, and by ionic defects at high P_{O_2}. In the case of TiO_2 doped with Nb_2O_5, the situation differed in the oxidized state in that the dopant itself was exsolved in the form of $TiNb_2O_7$[8]

$$xNb_{Ti}^{\cdot} + xe' + x/2\,Ti_{Ti} + 3xO_0 + x/4\,O_2 \underset{\text{reduction}}{\overset{\text{oxidation}}{\rightleftarrows}} x/2\,TiNb_2O_7 \qquad (8.9)$$

Both Eqs. (8. 8) and (8. 9) represent detailed defect versions of the basic phenomenon

$$2e^-(\text{crystal}) + 1/2\,O_2\,(\text{gas}) \underset{\text{reduction}}{\overset{\text{oxidation}}{\rightleftarrows}} O^=(\text{crystal}) \qquad (8.10)$$

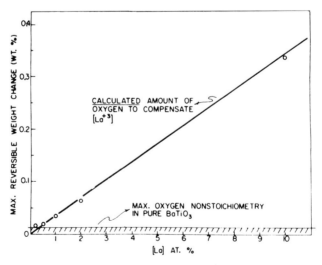

Fig. 8. 2 The reversible weight change of La_2O_3-doped $BaTiO_3$ between equilibrations in CO and O_2 at $1,060°C$ as a function of La content. [Reproduced from Ref. 8) with permission from the authors and North-Holland Publishing Co.]

This general behavior has now been confirmed for a number of similar cases for donor-doped, reduction-type oxides.[12,13]

Equilibrium measurements have since been made to show how the oxygen content of donor-doped samples varies with P_{O_2} and temperature. Figure 8.3 shows the equilibrium conductivity results of Daniels and Härdtl[14] for $BaTiO_3$ doped with La^{+3}. The sample with 0% La shows characteristic acceptor-doped behavior, due to naturally-occurring impurities, with a conductivity minimum separating oxygen-excess and oxygen-deficient regions. As the La-content increases, a distinct plateau develops in the region $10^{-10} < P_{O_2} < 10^{-5}$ atm, whose level increases with the amount of La^{+3}. This is the region in which the electron concentration is fixed by the donor-content, $n \approx (La_{Ba}^{\cdot})$. Figure 8.4 shows that the conductivity at the plateau is consistent with this equality with an electron mobility of 0.1 cm²/volt sec. The latter value is in good agreement with mobilities obtained from high temperature Hall measurements.[4] At very low P_{O_2} the conductivity rises again with decreasing P_{O_2} as the loss of oxygen according to Eq. (8.6) becomes the major source of defects. The conductivity drops rapidly with increasing P_{O_2} above 10^{-5} atm as oxygen enters the crystal according to Eq. (8.8) and consumes the electrons. The functional dependence was reported as $\sim P_{O_2}^{-1/3.7}$ in this region, close to the expected dependence of $P_{O_2}^{-1/4}$ when the donors are predominantly compensated by an ionic defect.[15,16]

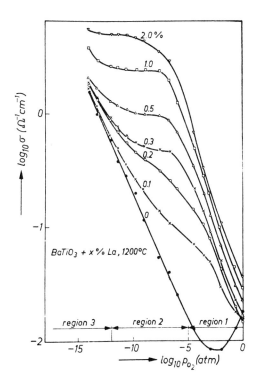

Fig. 8.3 The equilibrium conductivity of polycrystalline, La_2O_3-doped $BaTiO_3$ at 1,200°C as a function of P_{O_2} and La content. [Reproduced from Ref. 14) with permission from the authors and N. V. Philips.]

The weight change as a function of P_{O_2} for various $La_{Ba}^{\cdot}$ concentrations in $SrTiO_3$ is shown in Fig. 8.5.[15] The change is normalized to the weight at very low P_{O_2}, where it is assumed that the oxygen sublattice is filled in a single phase material having the composi-

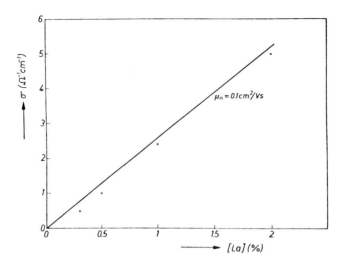

Fig. 8.4 Conductivity of La_2O_3-doped $BaTiO_3$, taken from the plateau region of Fig. 8.3, as a function of La-content. [Reproduced from Ref. 14) with permission from the authors and N. V. Philips.]

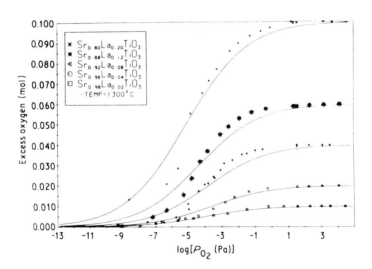

Fig. 8.5 The weight change at 1,300°C as a function of P_{O_2} for La_2O_3-doped $SrTiO_3$ plotted in terms of excess oxygen relative to the reduced state. [Reproduced from Ref. 15) with permission from the authors and Chapman and Hall.]

tion $Ba_{1-x}La_xTi\,e_x^-O_3$. With increasing P_{O_2}, the weight increases as oxygen enters the system and conbines with Ti to form V_{Ti}'''' and a Ti-rich second phase. The weight becomes independent of P_{O_2} at high P_{O_2} where $[La_{Ba}^{\cdot}] \approx 4[V_{Ti}''''']$. The total weight change corresponds to half a gram atom of oxygen for each gram atom of $[La_{Ba}^{\cdot}]$, as expected.

From the principles of defect chemistry, one can derive the relative dependence of the concentrations of electrons, holes, and ionic defects on P_{O_2} under equilibrium conditions, and this is shown schematically in Fig. 8.6. The trace of the electron concentration corresponds to the conductivity results shown in Fig. 8.3. At high P_{O_2}, where V_{Ti}'''' becomes the main compensating defect, the electron concentration decreases with increasing P_{O_2}. This is still not sufficient to give an insulating material, however, as long as $n > p$. At $P_{O_2} = 1$ atm the n-type conductivity has nearly declined to that of the conductivity minimum of acceptor-doped material (see Fig. 8.3) where the conductivity will become p-type with further increase in P_{O_2}. Therefore the hole concentration is shown rising toward the declining electron concentration as P_{O_2} increases. It must be concluded that additional oxidation occurs on cooling these materials in air toward room temperature, such that p becomes greater than n and the observed insulating behavior is thus achieved.

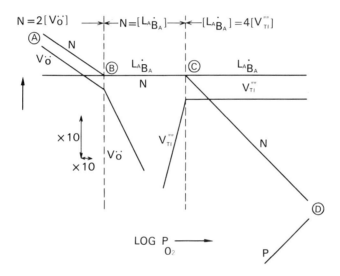

Fig. 8.6 Log-log plot of defect concentrations in La_2O_3-doped $BaTiO_3$ as a function of P_{O_2} at constant temperature. n and p are the electron and hole concentrations. Approximate conditions of change neutrality are shown for each region. Room temperature semiconduction is achieved for ideally quenched samples over the entire range of P_{O_2} since $n > p$ (line segment A-B-C-D).

The reduction-type oxides are semiconducting only when n-type. This can be achieved by reduction or by donor-doping, and the latter also requires the material to be reduced. Except in the latter cases, equilibration with an oxidizing ambient such as air will always yield an insulating material.

8.3.3 P-Type Semiconductors

The p-type, or oxidation-type, semiconductors are compounds containing a transition metal in its lowest oxidation state, e.g. NiO, MnO, Cr_2O_3, and $LaCrO_3$. The presence of oxidizable cations favors the addition of a stoichiometric excess of oxygen, and after equilibration in air, they are generally oxygen-excess, p-type semiconductors. Because the cation has a higher oxidation state, holes are easily tolerated ($Mn^{+2} + h^\cdot = Mn^{+3}$) and acceptor levels are very shallow. As a result, the holes remain free to conduct down to room temperature and below. When reduced to an oxygen-deficient stoichiometry, however, the n-type conductivity observed at high temperatures disappears on cooling as the electrons become trapped by the deep donor levels associated with the oxygen-deficient defects, usually $V_0^{\cdot\cdot}$

$$V_0^{\cdot\cdot} + 2e' \rightleftharpoons V_0 \tag{8.11}$$

The behavior is thus just the reverse of the reduction-type semiconductors.

When oxidation-type semiconductors are acceptor-doped, they have all the requirements necessary to be good semiconductors after equilibration in air. This includes an oxidizable species that favors a stoichiometric excess of oxygen, extrinsic oxygen vacancies to accommodate the excess oxygen, and a band structure that tolerates the presence of free holes. The intrinsic $V_0^{\cdot\cdot}$ result from the acceptor impurity as in the case of $LaCrO_3$ doped with MgO[17]

$$2\ MgO + Cr_2O_3 \longrightarrow 2\ Mg'_{La} + 2\ Cr_{Cr} + 5O_0 + V_0^{\cdot\cdot} \tag{8.12}$$

This corresponds to the composition under reducing conditions. At higher P_{O_2}, oxidation occurs according to Eq. (8.4) until the oxygen sublattice becomes saturated.

Figure 8.7 shows the reversible weight change observed for $La_{0.9}Mg_{0.1}CrO_3$ as a function of P_{O_2}.[17] As in the case of donor-doped reduction-type semiconductors, the total weight change between oxidized and reduced states corresponds to an amount of oxygen that is the chemical equivalent of the impurity states introduced, i.e. $x/2$ gram atoms of oxygen for $La_{1-x}Mg_xCrO_3$. The oxidation and reduction reactions are represented by Eq. (8.4). In the case of Fig. 8.7, the weight change is normalized to a filled oxygen sublattice that occurs at high P_{O_2} with the composition $La_{1-x}Mg_xCr\ h^\cdot_xO_3$, and the plot is in terms of increasing oxygen deficiency with decreasing P_{O_2}. Because of this difference in normalization, the weight changes in Figs. 8.5 and 8.7 are represented as oxygen excesses and deficiencies, respectively. Figure 8.7 also shows how the conductivity drops off with decreasing P_{O_2} as holes are consumed by the loss of oxygen.

Contrary to the case of donor-doped reduction-type semiconductors, the formation of an ionic defect to compensate the impurity is very simple in this case; it is the oxygen vacancy. Compensation for the acceptor impurity thus changes from $[Mg'_{Cr}] \approx p$ at high P_{O_2} to $[Mg'_{Cr}] \approx 2[V_0^{\cdot\cdot}]$ at low P_{O_2}. Figure 8.8 shows how the defect concentrations change with P_{O_2}.

Once the oxygen loss has achieved full compensation of the acceptor impurity by $V_0^{\cdot\cdot}$, further reduction will result in oxygen-deficient, n-type behavior according to Eq. (8.6). The material can then be expected to become an insulator on cooling as the electrons are trapped out as shown in Eq. (8.11).

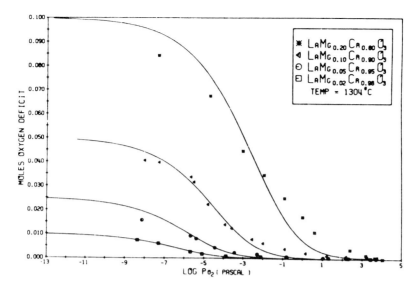

Fig. 8. 7 The weight change at 1,304°C as a function of P_{O_2} for MgO-doped LaCrO$_3$ plotted in terms of oxygen deficit relative to the oxidized state. [Reproduced from Ref. 17) with permission from the authors and the American Ceramic Society.]

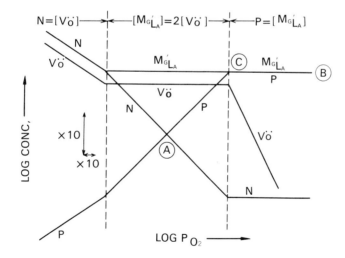

Fig. 8. 8 Log-log plot of defect concentrations in MgO-doped LaCrO$_3$ as a function of P_{O_2} at constant temperature. n and p are the electron and hole concentrations. Approximate conditions of charge neutrality are shown for each region. Room temperature semiconduction is achieved for ideally quenched samples for region where $p > n$ (line segment A–C–B).

8.3.4 Summary of Electronic Conduction

The behavior of oxidation-type semiconductors is seen to be the opposite of that of the reduction-type semiconductors. Semiconduction is achieved by equilibration in an oxidizing atmosphere, and the effect is enhanced by the presence of an acceptor impurity that provides room in the lattice for corresponding amounts of excess oxygen. Strong reduction or the addition of donor impurities leads to insulating behavior because excess electrons are effectively trapped at room temperature.

High levels of semiconduction in reduction-type compounds are achieved only in the reduced state, and reduction is enhanced by the presence of a donor impurity oxide that contains more oxygen than the binary component oxide it replaces. Loss of this excess oxygen leaves additional electrons in the lattice that enhance the conductivity.

In both Figs. 8. 6 and 8. 8, the P_{O_2} at which $n = p$ indicates the transition between semiconducting and insulating behavior. If the material is ideally quenched from the equilibration temperature, this boundary would be accurately defined. Since some compositional adjustment almost always occurs during cooling, the practical transition may be displaced somewhat from its ideal location.

8. 4 Ionic Conduction

8.4.1 General

Solid ionic conductors are primarily used as solid electrolytes in battery configurations for either the direct electrochemical generation of energy, as in the use of β-Al_2O_3 in the Na-S cell, or as an electrochemical sensor, such as the use of doped ZrO_2 in the oxygen concentration cell that analyzes the exhaust gases in automobiles. For such applications, the ionic conductivity must be in the same range as that of the more conventional electrolytes, aqueous solutions and fused salts. In the latter two systems conductivities up to about 1 ohm^{-1}cm^{-1} can be achieved and a number of solid systems are capable of approaching this level. At the same time, the electronic conductivity should be insignificant by comparison, and this is directly related to the chemical properties of the metallic constituent. If the cations in the solid electrolyte are easily reduced or oxidized to adjacent oxidation states, the band gap will be small and the electronic conductivity relatively high. Thus the useful electrolytes are compounds of main group metals, such as the Na and Al in β-Al_2O_3, or of metals of the second or third transition periods in their group oxidation state, such as Zr or Hf in acceptor-doped ZrO_2 or HfO_2. Such multivalent metals as Mn, Fe, and V find no application in solid electrolytes.

Ionic conductors are conveniently classified into three groups:

(a) Intrinsic Ionic Conductors

Conduction is due to ionic defects created by thermally activated ionic disorder. This very limited group includes the silver halides, the alkali halides, and the fluorite structure halides such as PbF_2 and BaF_2.

(b) Extrinsic Ionic Conductors

The charge carriers result from the impurity content, either naturally occurring or deliberately added. The most notable examples are the acceptor-doped fluorite-structure oxides, such as ZrO_2 or CeO_2 doped with CaO or Y_2O_3.

(c) Superionic Conductors

In these materials one of the sublattices becomes totally disordered, i.e. it melts, and its ions move readily through an open structure maintained by the other constituents. Well-known examples include the high temperature, cubic phase of AgI, and β-Al_2O_3. This is the extreme limit of intrinsic ionic conduction since impurities are not the cause of the high conductivities.

8.4.2 Intrinsic Ionic Conductors

The dominant intrinsic disorder in the silver halides is cation Frenkel disorder as shown in Eq. (8. 2). This reaction corresponds to the movement of a Ag^+ from its normal lattice site (an octahedral site in the cubic-close-packed anion array) to an interstitial (tetrahedral) site. In AgBr, the energetic cost of disorder is very modest, only about 1.2 eV, because the cation is monovalent and highly polarizable, which helps it to fit into the undersized interstitial site.[18] Approximately 1% of the cations are disordered in AgBr near the melting point of 432 °C, so only modest levels of purity are required for intrinsic disorder to be the major source of defects. The $Ag_i^{\cdot}$ moves very easily by a collinear interstitialcy mechanism, with an enthalpy of motion of only about 0.1 eV. The resulting high mobility, combined with the relatively high defect concentration, gives very respectable levels of conductivity, e.g. about 0.3 $ohm^{-1} cm^{-1}$ near the melting point. At a given temperature, AgI is somewhat more conducting and AgCl somewhat less. The size of the interstitial site, and the ease of ionic movement, both increase with the size of the anion.

In the alkali halides, Schottky disorder is preferred, as in the case of NaCl

$$nil \rightleftharpoons V'_{Na} + V^{\cdot}_{Cl} \qquad (8.13)$$

The enthalpy of defect formation is 2.45 eV for NaCl[19], about twice that of AgBr. Thus the defect concentration is only about 0.02% near the melting point of 801 °C. Moreover, the enthalpy of motion of the more mobile defect, the V'_{Na}, is 0.65 eV. Therefore the ionic conductivity of NaCl is only 10^{-3} $ohm^{-1}cm^{-1}$ near its melting point.

The alkaline earth fluorides are an intermediate case. The preferred type of disorder is of the anion Frenkel type

$$F_F + V_I \rightleftharpoons F'_I + V^{\cdot}_F \qquad (8.14)$$

This is favored because the anion is singly charged and the interstitial site (the octahedral site in an expanded face-centered-cubic array of cations) is larger than the normal (tetrahedral) anion site. It is also surrounded by cations making it a hospitable site for an interstitial anion. The enthalpy for defect formation has been reported to be 2.3 eV in SrF_2 and about 1.8 eV in BaF_2, while the enthalpies of motion are about 0.8 eV for both defects.[18] Near the melting point there is a dramatic increase in the defect concentration, changing the behavior to that of the superionic conductors.

All of these examples involve only singly charged defects, because the energetic cost of disorder increases rapidly with the charge on the defects. The enthalpy of formation of

Schottky defects in MgO, for example, has been calculated to be 7.8 eV,[20] much higher than the value of 2.45 eV for the isostructural NaCl. Thus there are no clear examples of intrinsic ionic conductors where the defects are more than singly charged.

8.4.3 Extrinsic Ionic Conductors

The limited number of intrinsic ionic conductors results from the low concentration of charge carriers caused by high enthalpies of disorder. Intrinsic disorder involves at least two defects and the enthalpy reflects the cost of forming the appropriate combination. One of the defect species may be much more favorable than the others and it is then possible to increase its concentration by addition of an aliovalent impurity. Useful conductivities can be achieved when the structure does not seriously restrict the mobility of the defect whose concentration is enhanced. Thus acceptor-doped oxides having the fluorite structure have ionic conductivities up to 0.1 ohm^{-1}cm^{-1} at 1,000 °C. The well-known case of ZrO_2 doped with CaO is a good example

$$CaO \longrightarrow Ca''_{Zr} + O_0 + V_0^{\cdot\cdot} \tag{8.15}$$

The replacement of ZrO_2 by CaO results in a corresponding number of oxygen vacancies. The solubilities of CaO, Y_2O_3, MgO, etc. in ZrO_2, HfO_2, ThO_2, and CeO_2 can be as much as 20%, indicating that $V_0^{\cdot\cdot}$ is a rather favorable defect in these compounds. The vacancy is also able to move quite easily through the open fluorite structure. The measured activation energy of conduction for Y_2O_3-doped ZrO_2 is about 1 eV and this should correspond to the enthalpy of motion for the oxygen vacancy.[21]

Significant ionic conduction has also been observed in acceptor-doped oxides having the perovskite structure, e.g. $CaTiO_3$, $SrTiO_3$, and $LaAlO_3$. The charge carrier is apparently the oxygen vacancy. Takahashi[21] has reported ionic transport numbers of 0.9–1.0 for such compositions as $La_{0.7}Ca_{0.3}AlO_{3-x}$, $CaTi_{0.7}Al_{0.3}O_{3-x}$, and $SrTi_{0.9}Al_{0.1}O_{3-x}$. The conductivities at 1,000 °C are in the range 10^{-2}–10^{-1} ohm^{-1}cm^{-1}, slightly lower than those of the acceptor-doped, fluorite-structure oxides. Equilibrium conductivity measurements have also shown evidence of substantial ionic conduction in $BaTiO_3$ doped with Al_2O_3 and CaO.[22,23] In both cases, the impurity substitutes for Ti, creating a corresponding number of $V_0^{\cdot\cdot}$. All of the Ti-based compounds suffer from increasing electronic conduction at very high and very low values of P_{O_2}. At high P_{O_2}, the presence of $V_0^{\cdot\cdot}$ facilitates the addition of a stoichiometric excess of oxygen, according to Eq. (8.4), with a resulting increase in p-type conduction. At low P_{O_2} the compounds become oxygen-deficient, according to Eq. (8.6), because of the reducibility of the Ti^{+4} ion, and this results in increasing n-type conduction.

8.4.4 Superionic Conductors

The accepted designation of this class of materials is rather unfortunate since they do not share the zero resistance characteristic of electronic superconductors. In fact they are more closely analogous to normal metals, in that the charge carrying species is totally disordered, but that the mobilities are still finite. This family represents the ultimate limit of the intrinsic ionic conductors, but is treated separately because the disorder is complete on one sublattice and the carrier concentration is not thermally activated.

The oldest known example of this class is α-AgI, the cubic form of AgI stable above 146 °C. In this structure the large I$^-$ ions assume a body-centered-cubic arrangement and the Ag$^+$ are essentially free to move through the intervening space. The conductivity of this phase is 1.3 ohm^{-1}cm^{-1}, independent of temperature, and actually decreases when the material melts. The ordered passages in the solid crystal permit higher mobilities than the total disorder characteristic of the melt.

Double salts of AgI form the basis of a diverse family of superionic conductors. These involve combination of AgI with large ionic species as in Ag$_2$S, Ag$_2$HgI$_4$ and RbAg$_4$I$_5$. These all involve a transition to a highly conducting phase at some characteristic temperature. The Ag$_4$I$_5^-$ series was discovered nearly 20 years ago,[25-27] and is remarkable for having both very low transition temperatures and perhaps the highest ionic conductivities at room temperature of any known materials. The transition temperatures for KAg$_4$I$_5$ and RbAg$_4$I$_5$ are -136 °C and -155 °C, respectively, and the latter has a conductivity of 0.21 ohm^{-1}cm^{-1} at 20 °C.[27] The unit cell contains 4RbAg$_4$I$_5$ and has 56 iodide tetrahedra that share faces such as to provide easy passage of Ag$^+$ between them.[28]

A number of halides of large divalent cations have very high ionic conductivities just below their melting points, and this is apparently due to a complete disordering of the anion sublattice. The conductivity of PbF$_2$, which has the fluorite structure above 300 °C, increases steadily up to a value of 2.5 ohm^{-1}cm^{-1} at about 550 °C,[29] at which point the anion sublattice appears to have become completely disordered. The conductivity then remains essentially constant up to and even beyond its melting point of 855 °C. There is no further change in conductivity on melting. BaCl$_2$ and SrBr$_2$ have abrupt phase transitions to highly conducting states just below their melting points, as shown in Fig. 8.9.[30] As in PbF$_2$, the

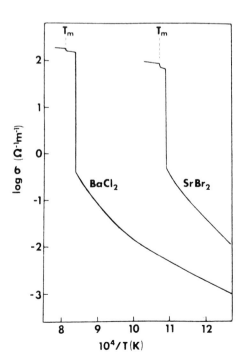

Fig. 8.9 The ionic conductivites of BaCl$_2$ and SrBr$_2$ as a function of temperature showing the abrupt increases in conductivity just below the melting points. [Reproduced from Ref. 30) with permission from the authors and Pergamon Press.]

highly conducting phase most probably has an anion-disordered fluorite structure. In $BaCl_2$ and $SrBr_2$, the entropy changes associated with the phase change, 14.4 and 12.1 J/K mol, respectively, are greater than the anomalously low entropies of fusion, 13.3 and 11.3 J/K mol.[30] It is clear that much of the entropy change usually associated with the decrease in order that results from melting has already occurred with the disordering of the anion sublattice in the solid state.

One of the most widely studied superionic conductors in recent years is β-Al_2O_3 and its close relation, β''-Al_2O_3. These have the compositions $Na_2O \cdot 11\,Al_2O_3$ and $Na_2O \cdot 5Al_2O_3$, respectively. These materials have been extensively described in a review by Kummer.[31] The structure of β-Al_2O_3 consists of two-dimensional spinel-like blocks containing only Al and O. The thickness of these blocks corresponds to four close-packed oxygen layers with Al^{+3} in both tetrahedral and octahedral sites. These blocks contribute nothing to the ionic conduction. The blocks are separated by layers, 11.27Å apart, that contain all of the Na^+ and an equal number of $O^=$. The $O^=$ serve both to bind the blocks together by means of Al-O-Al bonds, and to define the spacing between them. The Na^+ move freely in these sparsely populated planes and this results in a very high, two-dimensional conductivity. The ionic conductivity perpendicular to the Na^+-containing planes is immeasurably small. The conductivity of the polycrystalline material is 0.017 $ohm^{-1}cm^{-1}$ at 25 °C and 0.25 $ohm^{-1}cm^{-1}$ at 500 °C, and this is about one third of that of single crystals parallel to the conducting planes.[31] β''-Al_2O_3 is slightly more conducting because of its higher Na^+ content.

Na^+ can be easily replaced in β-Al_2O_3 by a wide variety of cations by means of simple ion-exchange. Na^+ seems to be the optimum choice for high conductivity, however. Replacement by either K^+ or Li^+, for example, gives lower conductivities. The former is too large to move easily in the conducting planes, and the latter is too small and tends to rest in off-center positions adjacent to the oxide ions of the spinel blocks.

8. 5 Conclusion

Compared with metals, electronically conducting ceramics have the advantages of thermal, chemical, and mechanical stability. Under severe operating conditions, these features may justify the use of ceramics in spite of their brittle, nonductile characteristics. If they are to be used, it is extremely important to understand how the conductivity of ceramic materials are affected by temperature and composition. This review has emphasized the differences between those oxides that are good conductors only after equilibration under reducing conditions, and those that require equilibration in an oxidizing atmosphere. Proper doping and equilibration can change the conductivities by many orders of magnitude.

Ceramic materials can have ionic conductivities equal to those of aqueous solutions and fused salts. In a few cases this is true even near room temperature, but more commonly the ceramics require high temperatures to achieve useful levels of conduction. The physical rigidity of a ceramic can make it useful where a liquid electrolyte would be impractical. The chemical inertness of ceramics makes them compatible with active materials as in the case of batteries, e.g. the Na-S battery in which the β-Al_2O_3 electrolyte is not severely attacked by the electrodes. High temperature stability can also be an advantage in electrochemical applications where electrode reactions are too slow at normal ambient temperatures.

Conducting ceramics are now used in a few applications, e.g. in $SrTiO_3$-based boundary-layer capacitors where the grains in the polycrystalline ceramic must be highly conducting, and in oxygen-activity sensors in automotive exhaust gases. They are being investigated for many more applications, and it can be safely predicted that their use will steadily increase as their properties are better understood, and their advantages are increasingly appreciated.

Acknowledgements

Support from the Divisions of Materials Research of the National Science Foundation and the Office of Naval Research for research related to the defect chemistry of oxides has been essential for the development of the concepts of this article and is gratefully acknowledged.

References

1) F. A. Kröger and H. J. Vink : in Solid State Physics, Vol. 3, ed. by F. Seitz and D. Turnbull, Academic Press, New York (1956) 307.
2) D. M. Smyth:"The role of impurities in insulating transition metal oxides," Prog. Solid State Chem., **15** (1984) 145–171.
3) D. M. Smyth : "Defects and transport in $LiNbO_3$," Ferroelectrics, **49** (1983) 419–428.
4) A. M. J. H. Seuter : "Defect chemistry and electrical transport properties of barium titanate," Philips Res. Rep., Suppl., 3 (1974) 84.
5) D. M. Smyth : "Compositional characteristics of dielectric oxides," to be published in the Proceedings of Thirty-First Sagamore Army Materials Research Conference (Aug. 1984).
6) G. H. Jonker and E. E. Havinga : "The influence of foreign ions on the crystal lattice of barium titanate," Mater. Res. Bull., **17** (1982) 345–350.
7) H. Chan, M. P. Harmer and D. M. Smyth: "Electron microscopy studies of Nb-doped $BaTiO_3$," American Ceramic Society, 86th Annual Meeting (May 1984).
8) N. G. Eror and D. M. Smyth : "Oxygen stoiochiometry of donor-doped $BaTiO_3$ and TiO_2," in the Chemistry of Extended Defects in Non-Metallic Solids, ed. by L. Eyring and M. O'Keeffe, North-Holland, Amsterdam (1970) 62–74.
9) H. M. O'Bryan, Jr. and J. Thomson, Jr. : "Phase equilibria in the TiO_2-rich region of the system BaO-TiO_2, " J. Am. Ceram. Soc., **57** (1974) 522–526.
10) T. Negas, R. S. Roth, H. S. Parker and D. Minor : "Subsolidus phase relations in the $BaTiO_3$-TiO_2 system," J. Solid State Chem., **9** (1974) 297–307.
11) R. K. Sharma, N.-H. Chan and D. M. Smyth: "Solubility of TiO_2 in $BaTiO_3$," J. Am. Ceram. Soc., **64** (1981) 448–451.
12) U. Balachandran and N. G. Eror : "Self-compensation in lanthanum-doped calcium titanate," J. Mater. Sci., **17** (1982) 1795–1800.
13) N. G. Eror and U. Balachandran : "Self-compensation in lanthanum-doped strontium titanate," J. Solid State Chem., **40** (1981) 85–91.
14) J. Daniels and K. H. Härdtl : "Electrical conductivity at high temperatures of donor-doped barium titanate ceramics," Philips Res. Rep., **31** (1976) 489–504.
15) B. F. Flandermeyer, A. K. Agarwal, H. U. Anderson and M. M. Nasrallah : "Oxidation-reduction behavior of La-doped $SrTiO_3$," J. Mater. Sci., **19** (1984) 2593–2598.
16) M. M. Nasrallah, H. U. Anderson, A. K. Agarwal and B. F. Flandermeyer : "Oxygen

activity dependence of the defect structure of La-doped $BaTiO_3$," J. Mater. Sci., **19** (1984) 3159-3165.

17) B. K. Flandermeyer, M. M. Nasrallah, A. K. Agarwal and H. U. Anderson : "Defect structure of Mg-doped $LaCrO_3$ model and thermogravimetric measurements," J. Am. Ceram. Soc., **67** (1984) 195-198.

18) A. D. Franklin : "Statistical thermodynamics of point defects in crystals," in Point Defects in Solids, Vol. 1, ed. by J. H. Crawford, Jr. and L. M. Slifkin, Plenum Publishing Corp., New York (1972) 1-101.

19) R. G. Fuller : "Ionic conductivity (including self-diffusion)," in Point Defects in Solids, Vol. 1, ed. by J. H. Crawford, Jr. and L. M. Slifkin, Plenum Publishing Corp., New York (1972) 103-150.

20) M. J. L. Sangster and D. K. Rowell : "Calculation of defect energies and volumes in some oxides," Phil. Mag. **A**, **44** (1981) 613-624.

21) T. Takahashi : "Solid electrolyte fuel cells (theoretics and experiments)," in Physics of Electrolytes, Vol. 2, ed. by J. Hladik, Academic Press, New York (1972) 989-1049.

22) N.-H. Chan, R. K. Sharma and D. M. Smyth : "Nonstoichiometry in acceptor-doped $BaTiO_3$," J. Am. Ceram. Soc., **65** (1982) 167-170.

23) J. Appleby, Y. H. Han and D. M. Smyth : "Calcium as an acceptor dopant in $BaTiO_3$," American Ceramic Society, 85th Annual Meeting (April 1983).

24) Y. Sakabe, J. Appleby, Y. H. Han, D. Wintergrass and D. M. Smyth : "Compensation of donor impurities in $BaTiO_3$ by calcium," American Ceramic Society, 86th Annual Meeting (May 1984).

25) J. N. Bradley and P. D. Greene : "Potassium iodide + silver iodide phase diagram, high ionic conductivity of KAg_4I_5," Trans. Faraday Soc., **62** (1966) 2069-2075.

26) J. N. Bradley and P. D. Greene : "Solids with high ionic conductivity in Group I halide systems," Trans. Faraday Soc., **63** (1967) 424-430.

27) B. B. Owens and G. R. Argue : "High-conductivity solid electrolytes: MAg_4I_5," Science, 157 (1967) 308-310.

28) S. Geller : "Crystal structure of the solid electrolyte, $RbAg_4I_5$," Science, 157 (1967) 310-312.

29) C. E. Derrington and M. O'Keeffe : "Anion conductivity and disorder in lead fluoride," Nature Phys. Sci., 246 (1973) 44-46.

30) C. E. Derrington and M. O'Keeffe : "The solid electrolyte behavior of barium chloride and strontium bromide," Solid State Commun., **15** (1974) 1175-1177.

31) J. T. Kummer : "β-alumina electrolytes," Prog. Solid State Chem., **7** (1972) 141-175.

32) N.-H. Chan and D. M. Smyth : "Defect chemistry of dopor-doped $BaTiO_3$," J. Am. Ceram. Soc., **67** (1984) 285-288.

9. Preparation and Application of Fine Powder from the Metal Alkoxide

Yoshiharu OZAKI*

Abstract

Utilization of metal alkoxides provides one of the most promising powder preparation methods in the manufacture of electroceramics for integrated circuits. Powder preparation from metal alkoxides and industrial applications for the alkoxy-derived powders are discussed in this paper. In addition, the practical application of the metal alkoxide process is discussed, specifically in a $BaTiO_3$ multilayer capacitor, $(Pb,La)(Zr,Ti)O_3$ (PLZT) switch, Al_2O_3 thin substrate and Al_2O_3 tweeter diaphram.

Keywords: fine powder, metal alkoxides, perovskite compound $BaTiO_3$, $YAlO_3$, valence compensated perovskite, multilayer capacitor, PLZT light switch, speaker diaphram.

9.1 Introduction

The memory capacity of dynamic RAMs is increasing year by year. Advances in IC and related technology have resulted in 1 micron-wide wiring allowing the manufacture of a 1 MB dynamic RAM. Along with this development has arisen the need for miniaturization of electronic devices. Ceramics are very promising for this purpose because of their properties allowing measurement and control of physical and chemical parameters using a microprocessor.

The miniaturization of electroceramic devices must be achieved without a decline in quality, reliability or cost performance. This may be achieved by the introduction of solid state and/or multilayer devices. Therefore, the development of microprocessing technology in the manufacture of electroceramics has recently attracted special interest. The desired dimensions for these electroceramic devices are below 10 μm for the multilayer capacitors and below 100 μm for multilayer substrates. However, it is difficult to obtain these thicknesses using conventional starting powders.

Conventional powders have a particle size of about 1 μm, and after firing they grow to several microns. Consequently conventional powder is too large to fabricate 10-100 μm ceramics devices with a good physical structure. In addition, conventional powders contain inhomogeneities in proportion to particle size. This compositional inhomogeneity is not a

---

* Department of Industrial Chemistry, Faculty of Engineering, Seikei University, 3-3-1, Kichijoji-kitamachi, Musashino, Tokyo 180, Japan.

serious problem in normal ceramics whose dimensions are large in comparison with particle size. In other words, if a dimension of the electroceramic components is large enough for the particle size of starting powders, the properties of electroceramics are not significantly affected by this local compositional variation.

In addition to the problems of particle size and compositional inhomogeneity of the starting powders, the particle size effect on a metallized conducting layers must be considered. Good continuity of metallized conducting layers largely depends on the roughness of the ceramic surface. A good smooth surface is obtained only by using fine starting powders.

Particle size distribution is also an important factor in preventing abnormal grain growth resulting from the discontinuity in the metallized conducting layer. Abnormal grain growth is caused mainly by wide variations in particle size in the starting powders. As already mentioned, the starting powders for advanced electroceramics are ultrafine, monodispersive and compositionally homogeneous on the particle scale.

9.2 Powder Preparation from Metal Alkoxides

Metal alkoxides have the basic formula $M(OR)_n$, where M is a metal, and R is an alkyl or an aryl. They are considered derivatives of alcohols (ROH) in which the hydroxylic hydrogen is replaced by metals M. Metal alkoxides are easily hydrolyzed and yield oxides, hydroxides and hydrated oxides in crystalline or amorphous form. The most convenient method for synthesizing metal alkoxides is the direct reaction of a metal with an alcohol:

$$M + nROH = M(OR)_n + n/2\ H_2$$

This direct reaction is limited to strong electropositive elements such as alkali and alkaline earth metals. Magnesium, beryllium, aluminum and lanthanons such as yttrium, scandium and ytterbium also react directly with alcohols, but in these cases a catalyst like mercuric chloride is necessary. Another convenient method for synthesizing metal alkoxides is the reaction of metal chloride with an alcohol in the presence of a base such as ammonia:

$$MCl_n + nROH + nNH_3 = M(OR)_n + nNH_4Cl$$

Fine powder preparation techniques from metal alkoxides are divided into two types. One is simple hydrolysis of the metal alkoxides. This is the fundamental method obtaining fine powders from metal alkoxides. The other is based on the utilization of sol fabricated from alkoxy-derived powders. Alkoxy-derived SiO_2 sol is especially useful for fine glass powder fabrication.

In the direct hydrolysis method, alkoxides are hydrolyzed directly to ceramic powders or ceramic precursor powders. The hydrolysis of simple alkoxides containing one metal atom in that molecule usually gives precipitates in hydroxide form, but sometimes in crystalline oxide form. For compounds containing two or more metal atoms, two types of alkoxide solutions are available for hydrolysis. One is a mixed solution of simple alkoxides which contain one metal atom in the alkoxide molecules as mentioned above. The other is a solution of double or more complex alkoxides which contain two or more metal atoms in the alkoxide molecules. In the former, properties of the solution are dominated by the constituent simple alkoxides. Thus, if constituent simple alkoxides have different hydrolysis rates, precipitates have a tendency to separate into hydrolysis products of the simple alkoxides. On the other hand, when a solution of double or more complex alkoxides is used, the difference in hydrolysis rates of constituent metal atoms may be regulated and relieved by the rapid stepwise hydrolysis reaction of the alkoxides.

Many compounds have been synthesized using these two types of alkoxide solutions. The results of this synthesis are summarized in Table 9.1 including the hydrolysis results of the simple alkoxides. Some precipitates are obtained as crystalline oxides. This group includes important basic ferroelectric or ferromagnetic compounds such as $BaTiO_3$, $SrTiO_3$, $BaZrO_3$, $MnFe_2O_4$, $NiFe_2O_4$ and $CoFe_2O_4$. In addition, solid solutions of crystalline precipitates such as $(Ba_{1-x}Sr_x)TiO_3$ and $(Mn_{1-x}Zn_x)Fe_2O_4$ are obtained in the crystalline form. Other precipitates are obtained in the form of hydroxides and hydrated oxides. For example, a precipitate of barium tin oxide $BaSnO_3$, barium tin hydroxide $BaSn(OH)_6$, is transformed to barium tin oxide by calcination. Most precipitates are obtained in the amorphous forms. They are transformed to oxides by calcination. Amorphous precipitates are divided into two types. One type is that in which crystallization of precipitates takes place in a single step at a crystallization temperature. This group includes $SrZrO_3$, $BaFe_{12}O_{19}$, $Pb(Zr_{1-x}Ti_x)O_3$, $(Pb_{1-x}La_x)(Zr_{1-y}Ti_y)O_3$, $Sr(Zn_{1/3}Nb_{2/3})O_3$, etc. The other type is that in which the crystallization of precipitates takes place through the formation of intermediates. This group contains $Pb(Mg_{1/3}Ta_{2/3})O_3$, $Sr(Mg_{1/3}Nb_{2/3})O_3$, etc. In this case powder synthesis proceeds as a solid state reaction. Consequently, there is no essential difference between this solid state reaction and that of the conventional method except the size of reaction particles.

Table 9.1 As-precipitated form of Alkoxy-derived powders.

Crystalline	$BaTiO_3$, $SrTiO_3$, $BaZrO_3$ $Ba(Ti_{1-x}Zr_x)O_3$, $Sr(Ti_{1-x}Zr_x)O_3$, $(Ba_{1-x}Sr_x)TiO_3$, $MnFe_2O_4$, $ZnFe_2O_4$, $(Mn_{1-x}Zn_x)Fe_2O_4$, Zn_2GeO_4, $PbWO_4$, $SrAs_2O_6$, $CoFe_2O_4$
Hydroxide or Hydrated Oxide	$BaSnO_3$, $SrSnO_3$ $PbSnO_3$, $CaSnO_3$, $MgSnO_3$, $SrGeO_3$, $PbGeO_3$, $SrTeO_3$
Amorphous (Synthesized without intermediate)	$SrZrO_3$, $Pb(Ti_{1-x}Zr_x)O_3$, $Pb_{1-x}La_x(Zr_yTi_{1-y})_{1-x/4}O_3$, $Sr(Zn_{1/3}Nb_{2/3})O_3$, $Ba(Zn_{1/3}Nb_{2/3})O_3$, $Sr(Zn_{1/3}Ta_{2/3})O_3$, $Ba(Zn_{1/3}Ta_{2/3})O_3$, $Sr(Fe_{1/2}Sb_{1/2})O_3$, $Ba(Fe_{1/2}Sb_{1/2})O_3$, $Sr(Co_{1/3}Sb_{2/3})O_3$, $Ba(Co_{1/3}Sb_{2/3})O_3$, $Sr(Ni_{1/3}Sb_{2/3})O_3$, $NiFe_2O_4$, $CuFe_2O_4$, $MgFe_2O_4$, $(Ni_{1-x}Zn_x)Fe_2O_4$, $(Co_{1-x}Zn_x)Fe_2O_4$, $BaFe_{12}O_{19}$, $SrFe_{12}O_{19}$, $PbFe_{12}O_{19}$, $R_3Fe_5O_{12}$ (R=Sm, Gd, Y, Eu, Tb), $Tb_3Al_5O_{12}$, $R_3Gd_5O_{12}$ (R=Sm, Gd, Y, Er), $RFeO_3$ (R=Sm, Y, La, Nd, Gd, Tb), $LaAlO_3$, $NdAlO_3$, $R_4Al_2O_9$ (R=Sm, Eu, Gd, Tb), $Co_3As_2O_8$

9.3 Example of Powder Synthesis from Metal Alkoxides

9.3.1 BaTiO₃

A flowsheet for the preparation of BaTiO$_3$ is shown in Fig. 9.1 as a typical example of powder preparation from metal alkoxides. Ba alkoxide can be prepared by the interaction of Ba metal and an alcohol under refluxing conditions. Ba alkoxide and Ti alkoxide are mixed and refluxed together in a solvent, such as a parent alcohol and benzene. This mixture is hydrolyzed to form crystalline BaTiO$_3$. In the same way, crystalline precipitates of SrTiO$_3$, BaZrO$_3$ are obtained. Solid solution of these compounds is also obtained by hydrolysis of a mixed solution of constituent alkoxides. The condition of precipitates varies with hydrolysis temperature. The X-ray diffraction patterns of precipitates hydrolyzed at room temperature are shown in Fig. 9.2. When the alkoxide solution is hydrolyzed at room temperature, the precipitate is amorphous. This amorphous precipitate is transformed to a mixed powder of BaCO$_3$ and TiO$_2$ by calcination at 780 °C for 7 hrs. Finally BaTiO$_3$ is synthesized by the solid-solid reaction. Therefore some problems remain in the compositional homogeneity on the particle level of resultant BaTiO$_3$ as already mentioned. In order to obtain chemically homogeneous BaTiO$_3$ from the alkoxide solution, metal alkoxides must be hydrolyzed at elevated temperatures. The X-ray patterns are shown in Fig. 9.3 of BaTiO$_3$ precipitates synthesized at the refluxing temperature of various alcohols. All BaTiO$_3$ was obtained in a single phase, and a crystalline precipitate was usually obtained at a higher refluxing temperature. An electron micrograph of as-precipitated and calcined BaTiO$_3$ is shown in Fig. 9.4. Precipitate is a fine, monodispersive powder with a grain size of several hundred angstroms. Electron micrographs of calcined powders show the important characteristic alkoxy-derived powders: significant grain growth does not occur at temperatures up to about 1,000 °C.

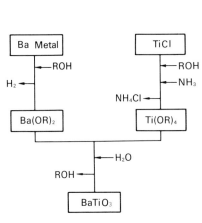

Fig. 9.1 Flowsheet of BaTiO$_3$ production.

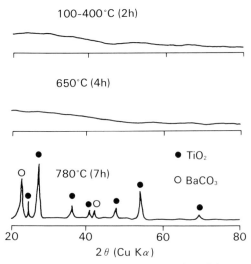

Fig. 9.2 X-ray diffraction patterns of precipitate at a various temperature. (Hydrolisis at R.T.)

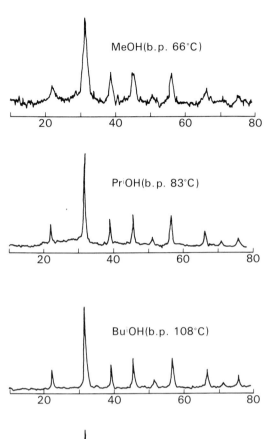

Fig. 9.3 X-ray diffraction patterns of BaTiO$_3$ precipitates which are synthesized at refluxing temperature of various alcohols.

Some solid solution precipitates are also easily obtained by hydrolysis of the mixed solution of simple alkoxides containing one metal atom per molecule. For example, the complex solid solution $(Ba_{1-x}Sr_x)(Ti_{1-y}Zr_y)O_3$ is synthesized by hydrolysis of a solution containing Ba alkoxide, Sr alkoxide, Ti alkoxide and Zr alkoxide in the mole ratio of $1-x : x : 1-y : y$. The lattice constants are shown in Fig. 9.5 of $Sr(Ti_{1-x}Zr_x)O_3$ solid solution precipitates as a function of the concentration of Zr. Lattice constants change linearly with Zr contents. This suggests that the continuous substitution of the Zr atom for the Ti atom is achieved by hydrolysis of the mixed solution of each constituent alkoxide.

Synthesis of alkali earth metal alkoxides can usually be prepared by the interaction of alkali earth metals and alcohols. This method is very easy, and is used frequently in

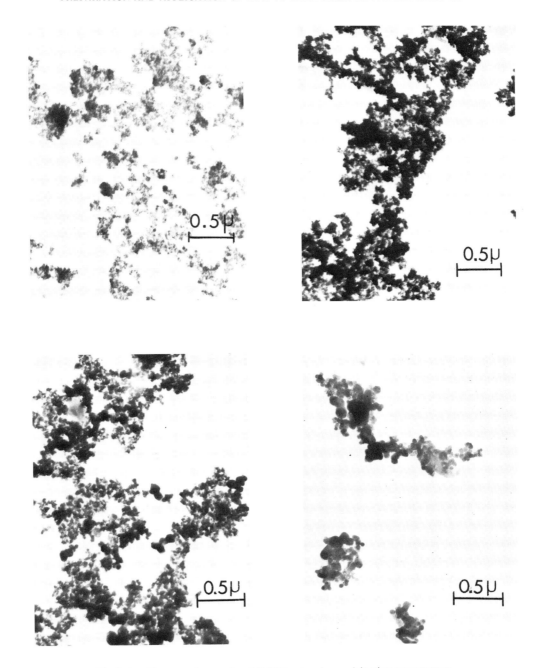

Fig. 9.4 Electron-micrographs of BaTiO$_3$ at various calcination temperatures.

laboratory experiments. However, alkali earth metals are expensive and unstable materials. Thus the development of low cost synthesis methods for alkali earth titanates and zirconates not using alkali earth metals as direct starting material is desired for industrial applications.

One such method is the reaction of the double alkoxide of the alkali metal titanium with an inorganic or organic salt solution of alkali earth metals. For example, crystalline BaTiO$_3$ and SrTiO$_3$ are synthesized by the reaction of sodium titanium alkoxide with

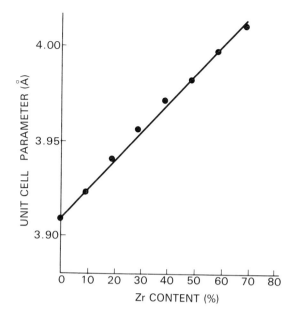

Fig. 9.5 Variation of unit cell parameter with composition for as-precipitated Sr$(Ti_{1-x}Zr_x)O_3$.

chlorides, acetates and oxalates of Ba and Sr. The results of X-ray diffraction and thermal analysis are shown in Figs. 9.6 and 9.7 of an as-precipitated BaTiO$_3$ synthesized by the reaction of sodium titanium alkoxide with Ba(CH$_3$COO)$_2$. The precipitate crystallizes well and the thermal analysis curves are similar to those for an alkoxy-derived powder.

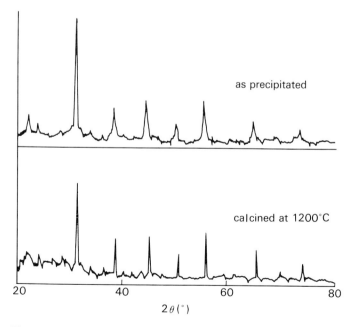

Fig. 9.6 X-ray diffraction pattern of BaTiO$_3$ synthesized by the reaction of alkoxides and Ba(CH$_3$COO)$_2$.

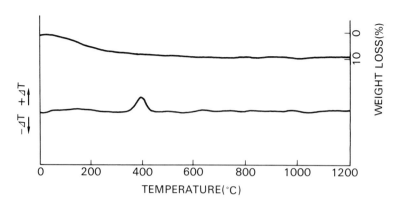

Fig. 9.7 Thermal analysis of BaTiO$_3$ synthesized by the reaction of alkoxides and Ba(CH$_3$COO)$_2$.

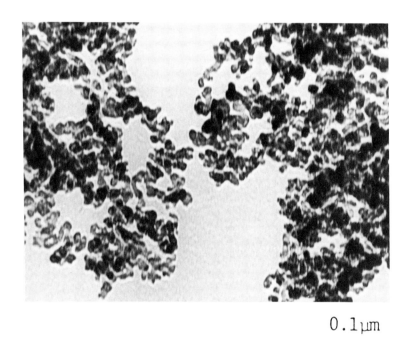

Sr	Ti	Sr	Ti
0.985	1.015	0.971	1.029
1.057	0.943	1.045	0.955
0.944	1.006	0.927	1.073
0.950	1.050	1.033	0.967
1.038	0.962	1.065	0.935
1.040	0.960	1.046	0.954
1.000	1.000	1.034	0.966
1.030	0.970	1.062	0.938
0.903	1.097	1.057	0.943
0.971	1.029	1.009	0.991
1.007	0.993	0.891	1.109
0.967	1.033	0.949	1.051

Fig. 9.8 Electron micrograph and results of quantitative analysis of alkoxy-derived SrTiO$_3$ particles. (TEMSCAN-100CX)

The advantage of powder preparation using metal alkoxides is that an ultrafine, monodispersive and compositionally homogeneous powder is obtained by a simple hydrolysis reaction. An electron micrograph and quantitative energy dispersive X-ray analysis are shown in Fig. 9.8 of alkoxy-derived as-precipitated $SrTiO_3$. As an example, the values for 24 analysis points are listed. As is obvious from the figure, the precipitate has very fine grain size in the 5 to 20 μm range, and has good compositional homogeneity.

9.3.2 $YAlO_3$

Perovskite structure $YAlO_3$ is considered to be in a metastable phase under atomospheric pressure. It is not synthesized in a pure phase by such ordinary solid state reactions as the reaction of aluminum hydrate with yttrium hydrate. Perovskite structure $YAlO_3$ has been synthesized in pure phase only by crystallization from melt at 2,400 °C. When the mixed solution of $Y(OR)_3$ and $Al(OR)_3$ is hydrolyzed, the perovskite structure $YAlO_3$ is not

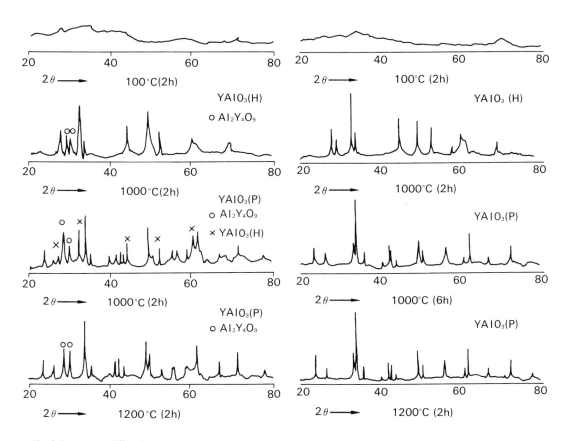

Fig. 9.9 X-ray diffraction pattern of the $YAlO_3$ composition precipitate calcined at various temperatures. (Starting materials are $Y(i-OC_3H_7)_3$ and $Al(i-OC_3H_7)_3$.)

Fig. 9.10 X-ray diffraction patterns of the $YAlO_3$ composition precipitate calcined at various temperatures. (Starting materials are $Y(Al(i-OC_3H_7)_4)_3$ and $Y(i-OC_3H_7)_3$.)

obtained by calcination at up to 1,200 °C. However, when the mixed solution of Y[Al(OR)$_4$]$_3$ and Y(OR)$_3$ is refluxed and hydrolyzed, the pure perovskite structure YAlO$_3$ is synthesized at 1,000 °C.

The results for these experiments are shown in Figs. 9.9 and 9.10. These results make it clear that complete mixing of Y atoms and Al atoms is achieved in the solution when Y[Al(OR)$_4$]$_3$ is used as a starting material. Al(OR)$_3$ and Y(OR)$_3$ form a rigid tetrameric structure in solution, which is not easily broken down. As a result, Y(OR)$_3$ and Al(OR)$_3$ do not mix completely at the molecular level, forming perovskite structure YAlO$_3$. Y[Al(OR)$_4$]$_3$ is less rigid than that of tetrameric Y(OR)$_3$ or Al(OR)$_3$. Both Y[Al(OR)$_4$]$_3$ and Y(OR)$_3$ react readily, and the complete mixing of Y atoms and Al atoms is achieved. Thus, the perovskite structure YAlO$_3$ is obtained in a metastable state at low temperature. This example makes it clear that the structure of molecular species in solution is also important for ceramic powder preparation from metal alkoxides.

9.3.3 A(B$_{1/3}$Sb$_{2/3}$)O$_3$ (A = Sr, Ba ; B = Co, Ni)

Some metal alkoxides are insoluble in a parent alcohol and other organic solvents. Compositionally homogeneous precipitates are not synthesized in a heterogeneous reaction system. Therefore in such systems it is necessary to liquefy these insoluble alkoxides. Sometimes this is achieved by the formation of double alkoxides prepared by the reaction of insoluble alkoxides with other constituent alkoxides. Valence compensated A(B$_{1/3}$Sb$_{2/3}$)O$_3$ is an example, where A is Sr or Ba, and B is divalent Co or Ni. The constituent alkoxide of this valence-compensated perovskite is Sr or Ba alkoxide, Co or Ni alkoxide and Sb alkoxide. In this case, Co and Ni simple alkoxides are not soluble by any common solvent of the other alkoxides. However, these alkoxides are liquefied by making a double alkoxide of Co and Sb or Ni and Sb, thus achieving uniformity in the reaction systems. For precipitate formation, alkoxide solution is usually hydrolyzed in an excess amount of water.

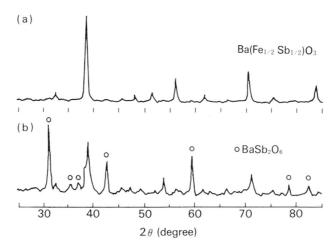

Fig. 9.11 X-ray diffraction patterns for hydrolysis products of Ba(Fe$_{1/2}$Sb$_{1/2}$)O$_3$ composition alkoxide solution calcined at 800 °C, 1.5 h. (a) using stoichiometric amount of water, and (b) excess amount of water.

Sometimes, however, the alkoxide solution must be hydrolyzed by a restricted amount of water in order to obtain a good precipitate. $Ba(Fe_{1/2}Sb_{1/2})O_3$ composition alkoxide is hydrolyzed into amorphous precipitate, which is transformed to oxide powders by calcination. When the alkoxide solution is hydrolyzed by the excess amount of water, an impurity phase $BaSb_2O_6$ appears. Single phase $Ba(Fe_{1/2}Sb_{1/2})O_3$ is obtained when an alkoxide solution is hydrolyzed by only the stoichiometric amount of water, as shown in Fig. 9. 11. The addition of the stoichiometric amount of water is performed by using the phase separation of the alcohol-water solution systems, for example, the $BuOH-H_2O$ system.

9. 4 Micro and Macro Structure Design of Ceramics by the Alkoxy-derived Powders

A block diagram of ceramic fabrication from metal alkoxides is shown in Fig. 9. 12. The fundamental structure design of ceramics by the alkoxy-derived powder is based on the direct use of these as-precipitated and calcined powders. Alkoxides are synthesized by the reaction of a metal or a metal chloride with an alcohol. Then alkoxides are hydrolyzed to ceramic powders or ceramic precursor powders. These reactions proceed at temperatures ranging from room temperature to about 100 °C When the hydrolysis products are oxides, these powders are available for the conventional fabrication process. Especially when an oxide powder is uesd for the fabrication of the plastic-ceramic composite, electroceramics can be obtained without sintering. On the other hand, if the hydrolysis product is dispersed easily in a liquid-vehicle, shaping processes utilizing a fluidized powder (slip, paste or sol) are available for the manufacture of ceramic thin films, fibers, filaments and small spheres. These fluidized materials are also available for preconsolidation of alkoxy-derived ultrafine

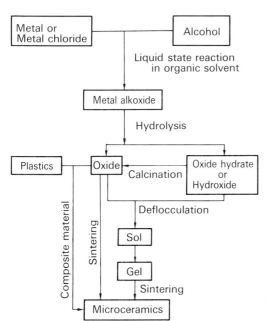

Fig. 9. 12 Block diagram for ceramic fabrication from metal alkoxide.

powders. Preconsolidation from fluidized starting materials is expected to play an especially important part in microstructure design of electroceramics in the future.

The advanced structure design of ceramics by the alkoxy-derived powders is characterized as follows, shown schematically in Fig. 9. 13.

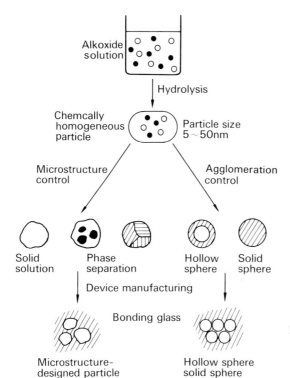

Fig. 9. 13 Micro and macrostructure design of powder particle and agglomeration by metal alkoxide.

(1) The microstructure of the powder particles is designed in situ on the particle level.
(2) The macrostructure is designed by packing the submicron spheres in the same way atoms pack together to form crystals.

Hydrolysis of metal alkoxides in most case gives chemically homogeneous powders on the particle level. Powder particles obtained by hydrolysis have the same cation ratio as the starting solution. When these powders are heat treated, each particle grows and crystallizes, achieving a stable microstructure according to their chemical and physical properties. In some case a solid solution is formed, while in others a phase separation occurs.

The microstructure design of powder particles by such techniques is based on a change from homogeneous solid to heterogeneous solid. In other words this is just an in situ phase change. However, powder preparation by ordinary solid state reaction is based on a contrary change. This in situ microstructure design of powder particles using the alkoxy-derived powders may provide a new ceramic manufacturing technique. These ceramic powders have optimum heat treatment properties.

To develop electronic devices using these powders, a suitable bonding material is required. An alkoxy-derived glass powder is very useful for this purpose. The macrostructure design of electroceramics from metal alkoxides is based on the utilization of the microsphere preconsolidated from fluidized starting materials. Using an alkoxy-derived ultrafine powder results in a well-dispersed and fluidized powder-vehicle system, very useful for the manufacture of the microspheres. Advanced ceramics with regular three-dimensional structures resembling single crystals are obtained by applying a layer by layer stacking technique developed to fabricate precious man-made opal. The layer by layer stacking of microspheres having different feromagnetic, feroelectric, piezoelectric and electroconductive properties may provide exotic, extremely anisotropic and functional ceramics.

9.5 Practical Application

9.5.1 Multilayer Capacitor

One of the most important products in the electroceramic industry is the capacitor. With advances in IC and LSI technology, consumption of many electric circuit parts for electrical products has decreased gradually, but consumption of multilayer capacitors is increasing especially fast. This is, because utilizing multilayer capacitors in additional electrical circuits for high performance and multifunction electrical products is essential to achieve compactness. $BaTiO_3$ is widely used for multilayer capacitors with many dopants added in practical applications to control their properties and the sintering process.

Capacitance is determined by the effective area and distance between electrodes. Large capacitance in multilayer capacitors is achieved by piling up dielectric layers of several ten μm thickness alternatelly with electrodes several μm thick. Variations in capacitance value increase in reverse proportion to dielectric layer thickness. These variations are a serious problem since the particle size of starting powders is comparable to

Table 9.2 Characteristics of alkoxy-derived multilayer capacitor.

Layer Thickness (μm)	Maximum Grain Size (μm)	Breakdown Voltage (V)	Life Index (Fit number)
30	2.0	1,200	0.4
	3.2	1,200	0.4
	4.2	1,100	0.6
	5.1	1,100	1.0
	6.0	950	1.2
	6.5	860	1.8
	7.7	750	2.0
20	2.0	860	0.8
	3.0	790	1.0
	4.0	640	1.2
	4.5	610	1.8
	5.0	540	3.2
10	1.6	460	1.2
	2.0	400	1.8
	2.8	300	3.6

dielectric layer thickness. Because electroceramic starting powders contain many additives in small amounts, it is difficult to disperse uniformly such additives in the starting powders.

In general, a conventional powder's compositional inhomogeneity increases in direct proportion to particle size. On the other hand, alkoxy-derived such powders are homogeneous on the particle level. In other words all particles in the starting powders are designed to have the same chemical composition. We and Mitsubishi Mining and Cement Co., Ltd. have developed a high quality, highly reliable multilayer capacitor using alkoxy-derived powders. Table 9. 2 shows the breakdown voltage and life index at a high temperature as a function of the maximum grain size at constant layer thickness for alkoxy-derived multilayer capacitors. The results shown are as follows:
(1) The high breakdown voltage and low failure rate (small Fit number) are attained with fine grain values.
(2) As the layer thickness decreases, so dose the breakdown voltage, while the rate of failure increases.
(3) For example, the maximum grain size, which assures a 1.8 in fit number is 6.5, 4.5 and 2.0 μm for layer thicknesses of 30, 20 and 10 μm, respectively. The thinner the layer thickness, the smaller the grain size necessary to maintain the same failure rate.

These results make it clear that larger capacities in products of the same size, i.e., thinner dielectric layers are achieved using starting powders with smaller particle size.

9. 5. 2 PLZT Light Switch

There are many electro-optical devices which utilize transparent PLZT ceramics. In such devices the intensity variations in transmitted light through each window is due to the local inhomogeneities in the ceramic microstructure. This local inhomogeneity is induced by the grain boundary and compositional inhomogeneity in the ceramic body. The grain boundary effect is greater when the width of the light window approaches the grain size of

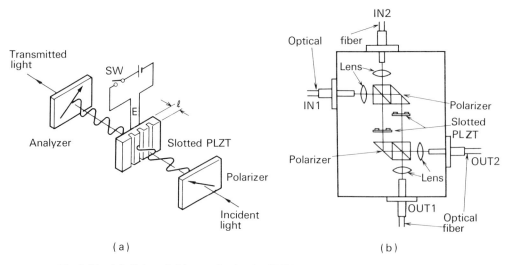

Fig. 9. 14 (a) light switching mechanism by PLZT, and (b) the application for 2 × 2 light switch.

PLZT. These problems can be avoided by utilizing fine grain and chemically homogeneous PLZT. We have synthesized PLZT powder from the metal alkoxide.

An akoxy-derived powder is very useful for the fabrication of this PLZT. A 2 × 2 light switch using alkoxy-derived PLZT shown schematically in Fig. 9. 14, has been produced by Yokogawa Hokushin Electric Corporation. Incident light is divided into two orthogonally polarized beams by a beam splitter, and then passes through PLZT. The light is polarized by PLZT, its output direction is determined, and then the light is recombined at the desired output. The light transmittance in relation to operating voltage is shown in Fig. 9. 15. The main characteristics of this light switch are an insertion loss of 2.5 dB, a switching time of 20 μs and an extinction ratio of 20 dB.

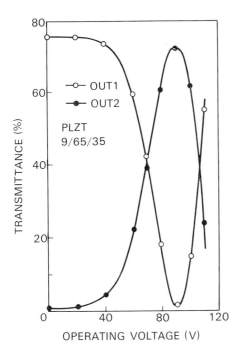

Fig. 9.15 Variation of transmittance with operating voltage of alkoxy-derived PLZT in a prototype.

9.5.3 Speaker Diaphram

The fundamental requirements for materials in a speaker diaphram are low density and high stiffness. Generally papers, plastics and metals are used widely because they are light and easy to work with. They are sometimes reinforced with ceramic or metal fibers to improve their stiffness. Ceramics are very light and have high stiffness, but they are difficult to fabricate as a very thin film with complex shapes. Conventional powders have a wide particle size distribution and a strong, irregular agglomaration which can't be broken down by grinding. These induce warpage and distortion in thin films. These phenomena appear critical in films less than 100 μm thick. However, good thin films of such thickness are obtained by using the alkoxy-derived powder. We and Mitsubishi Mining and Cement Co., Ltd. have been developed a technology for thin film fabrication.

A precursor film of alumina ceramics is made from alkoxy-delived behomite sol.

After forming it into the desired shape, it is sintered and transformed into an alumina ceramic. A dome-like alumina thin film was incorporated into a tweeter diaphram by JVC. The frequency response characteristics of this tweeter are shown in Fig. 9. 16 in comparison with a conventional tweeter with an aluminum diaphram. The high frequency limit of the ceramic diaphram is 42 kHz. This is about 2 times higher than the conventional metal diaphram. The ceramic diaphram thus provides a tweeter with a wider dynamic range and lower distortion. This is due to the high stiffness of the ceramic diaphram. The deformation measurements of the tweeter diaphrams by a holography. An aluminum diaphram deforms strongly at 42 kHz. But an alumina diaphram hardly deforms even at 55 kHz and shows an ideal piston motion at this frequency.

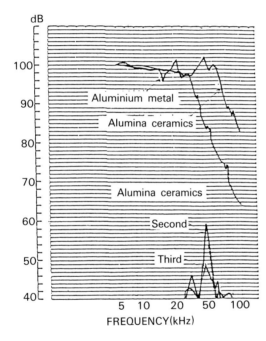

Fig. 9. 16 Frequency response characteristic of tweeter with alumina ceramic diaphram.

9. 6 Conclusion

This paper presented new powder preparation methods for the manufacture of electroceramics for electronics industry use. Powder preparation from metal alkoxides and industrial applications for the alkoxy-derived powders were also discussed. In addition, practical applications of the metal alkoxide processes were presented.

References

1) Y. Ozaki and M. Hideshima: "The production of transparent ceramic thin plates by sol-gel method," Proc. 52nd Annual Meeting of the Ceramic Society of Japan (1977) 25.

2) Y. Ozaki and M. Hideshima: "Production of transparent alumina thin plates," Proc. 14th Meeting on Powders (1976) 43.
3) Y. Ozaki and M. Hideshima: "Production of transparent TiO_2 thin plates from sols," Proc. 53rd Annual Meeting of the Ceramic Society of Japan (1978) 43.
4) Y. Ozaki and M. Hideshima: "Low temperature synthesis of transparent spinel," Proc. 54th Annual Meeting of the Ceramic Society of Japan (1979) 108.
5) Y. Ozaki, M. Hideshima and Y. Kojima: "Production of micro corundum balls using phase separation," Proc. 54th Annual Meeting of the Ceramic Society of Japan (1979) 95.
6) Y. Ozaki: "Production of functional masterials from metal alkoxides," Proc. 55th Annual Meeting of the Ceramic Society of Japan (1980) 127.
7) Y. Ozaki and Y. Kojima: "Synthesis of yttrium oxides from sols," Proc. 55th Annual Meeting of the Ceramic Society of Japan (1980) 126.
8) Y. Ozaki, M. Hideshima, Y. Kojima and T. Suzuki: "Production of $MgAl_2O_4$ spheres by sol-gel method," Proc. 55th Annual Meeting of the Ceramic Society of Japan (1980) 125.
9) Y. Ozaki, M. Hideshima, Y. Kojima and M. Shinoda: "Production of transparent mullite from metal alkoxides," Proc. 55th Annual Meeting of the Ceramic Society of Japan (1989) 124.
10) Y. Ozaki, M. Hideshima and Y. Shinohara: "Synthesis of PZT powders from metal from metal alkoxides," Proc. 55th Annual Meeting of the Ceramic Society of Japan (1980) 123.
11) Y. Ozaki, M. Hideshima and Y. Shinohara: "Synthesis of PZT powders from metal alkoxides," Proc. 55th Annual Meeting of the Ceramic Society of Japan (1980) 122.
12) Y. Ozaki, H. Ono and Y. Ito: "Synthesis of $MTiO_3$ from metal alkoxides," Proc. 56th Annual Meeting of the Ceramic Society of Japan (1981).
13) Y. Ozaki, Y. Shinohara and T. Kumazawa: "Synthesis of PLZT powders from metal alkoxides," Proc. 56th Annual Meeting of the Ceramic Society of Japan (1981).
14) Y. Ozaki, H. Ono and N. Takeshita: "Synthesis of Y_2O_3-Al_2O_3 system ceremics from metal alkoxides(II)," Proc. 56th Annual Meeting of the Ceramic Society of Japan (1981).
15) Y. Ozaki and J. Iwanari: "Synthesis of $(Ba,Sr)(Ti,Zr)O_3$ powders from metal alkoxides," Proc. 56th Annual Meeting of the Ceramic Society of Japan (1981).
16) Y. Ozaki and Y. Shinohara: "Synthesis of $MSnO_3$ powders from metal alkoxides," Proc. 56th Annual Meeting of the Ceramic Society of Japan (1981).
17) Y. Ozaki, H. Ono and N. Arakawa: "New preparation method for transparent alumina," Proc. 56th Annual Meeting of the Ceramic Society of Japan (1981).
18) Y. Ozaki, H. Ono and K. Onabuta: "Synthesis of Y_2O_3-Al_2O_3 system ceramics from metal alkoxides (I)," Proc. 56th Annual Meeting of the Ceramic Society of Japan (1981).
19) Y. Ozaki and Y. Shinohara: "Synthesis of MFe_2O_4 from metal alkoxides (M = Co, Ni)," Proc. 57th Annual Meeting of the Ceramic Society of Japan (1982).
20) Y. Ozaki and S. Mitachi: "Synthesis of MFe_2O_4 from metal alkoxides (M = Zn, Cu, Mg)," Proc. 57th Annual Meeting of the Ceramic Society of Japan (1982).
21) Y. Ozaki, Y. Shinohara, S. Mitachi and H. Morijiri: "Synthesis of indium double oxides from metal alkoxides," Proc. 57th Annual Meeting of the Ceramic Society of Japan (1982).
22) Y. Ozaki, Y. Shinohara, S. Mitachi and S. Hasegawa: "Synthesis of $RFeO_3$ and $R_3Fe_5O_{12}$ from metal alkoxides (R = rare earth elements)," Proc. 57th Annual Meeting of the Ceramic Society of Japan (1982).
23) Y. Ozaki: "Problems of ceramics production using metal alkoxides," Proc. 57th Annual Meeting of the Ceramic Society of Japan (1982).
24) Y. Ozaki and J. Iwanari: "New synthesis method of $(Ba,Sr)(Ti,Zr)O_3$ from metal alkoxides," Proc. 57th Annual Meeting of the Ceramic Society of Japan (1982).
25) Y. Ozaki, Y. Shinohara, S. Mitachi and H. Kobayashi: "Synthesis of valence compensated type Perovskites from metal alkoxides," Proc. 57th Annual Meeting of the Ceramic Society of Japan (1982).

26) Y. Ozaki, Y. Shinohara, S. Mitachi and H. Yoshida: "Synthesis of $LnAlO_3$ and $Ln_3Al_5O_{12}$ from metal alkoxides (Ln= Lanthanides)," Proc. 57th Annual Meeting of the Ceramic Society of Japan (1982).

27) Y. Ozaki, Y. Shinohara, S. Mitachi and M. Tanaka: "Synthesis of germanium double oxides from metal alkoxides," Proc. 57th Annual Meeting of the Ceramic Society of Japan (1982).

28) Y. Ozaki, Y. Shinohara and S. Mitachi: "Magnetization properties of magnetic materials prepared from metal alkoxides," Proc. 57th Annual Meeting of the Ceramic Society of Japan (1982).

29) Y. Ozaki, Y. Shinohara and M. Gohei: "Synthesis of $Ln_3Ga_5O_{12}$ from metal alkoxides," Proc. 58th Annul Meeting of the Ceramic Society of Japan (1983) 145.

30) Y. Ozaki, Y. Shinohara and K. Hamada: "Synthesis of tellurium double oxides from metal alkoxides," Proc. 58th Annual Meeting of the Ceramic Society of Japan (1983) 143.

31) Y. Ozaki and Y. Shinohara: "Synthesis of Mn-ferrite from metal alkoxides." Proc. 58th Annual Meeting of the Ceramic Society of Japan (1983) 155.

32) Y. Ozaki, Y. Shinohara and K. Yamamoto: "Synthesis of magnetoplumbite $MFe_{12}O_{19}$ (M = Sr, Ba, Pb) from metal alkoxides," Proc. 58th Annual Meeting of the Ceramic Society of Japan (1983)147.

33) Y. Ozaki, Y. Shinohara and J. Iwanari: "Synthesis of $M^{2+}TiO_3$ by the reaction of metal alkoxides and salt of organic acid," Proc. 58th Annual Meeting of the Ceramic Society of Japan (1983)153.

34) Y. Ozaki, Y. Shinohara and M. Wadasako: "Synthesis of Perovskite compounds containing Sb^{5+} from metal alkoxides," Proc. 58th Annual Meeting of the Ceramic Society of Japan (1983) 151.

35) Y. Ozaki, Y. Shinohara and H. Yoshida: "Ceramic powder preparation by spray-hydrolysis of metal alkoxides," Proc. 58th Annual Meeting of the Ceramic Society of Japan (1983) 149.

36) Y. Ozaki, Y. Shinohara and T. Ono: "Synthesis of tungsten double oxides from metal alkoxides," Proc. 58th Annual Meeting of the Ceramic Society of Japan (1983) 141.

37) Y. Ozaki: "Design of ceramic powder particle using metal alkoxides," Proc. 59th Annual Meeting of the Ceramic Society of Japan (1984) 237.

38) Y. Ozaki, M. Wadasako and M. Kubota: "Synthesis of Bi-contained double oxides from metal alköxides," Proc. 59th Annual Meeting of the Ceramic Society of Japan (1984) 227.

39) Y. Ozaki, K. Kaneko and S. Mitachi: "Rare earth oxides stabilized zironia derived from metal alkoxides," Proc. 59th Annual Meeting of the Ceramic Society of Japan (1984) 235.

40) Y. Ozaki and H. Yoshida: "(Pb, Sr)(Zr, Ti) O_3 powder preparation by spray-hydrolysis from metal alkoxides," Proc. 59th Annual Meeting of the Ceramic Society of Japan (1984) 223.

41) Y. Ozaki, M. Wadasako and S. Yamanaka: "Synthesis of ferrite from Fe (II) alkoxide," Proc. 59th Annual Meeting of the Ceramic Society of Japan (1984) 229.

42) Y. Ozaki, M. Wadasako and K. Hayashi: "Synthesis of low temperature glass from metal alkoxides," Proc. 59th Annual Meeting of the Ceramic Society of Japan (1984) 225.

43) Y. Ozaki, M. Wadasako and S. Okada: "Beryllium oxide powder preparation from metal alkoxides," Proc. 59th Annual Meeting of the Ceramic Society of Japan (1984) 231.

44) Y. Ozaki, M. Wadasako and Y. Kawase: "Synthesis of arsenic double oxides from metal alkoxides," Proc. 59th Annual Meeting of the Ceramic Society of Japan (1984) 233.

45) Y. Ozaki, K. Hayashi and H. Kawaharada: "Synthesis of ultrafine ruthenium oxide powders from ruthenium alkoxides," Proc. 60th Annual Meeting of the Ceramic Society of Japan (1985) 367.

46) Y. Ozaki and M. Wadasako: "Synthesis of hexaferrites from Fe(II) alkoxide," Proc. 60th Annual Meeting of the Ceramic Society of Japan (1985) 369.

47) Y. Ozaki, M. Wadasako and K. Iida: "Synthesis of Iron oxides from Fe(II) alkoxide," Proc. 60th Annual Meeting of the Ceramic Society of Japan (1985) 371.

48) Y. Ozaki, K. Hayashi and S. Yamamoto: "Solubility measurements of barium alkoxides", Proc. 60th Annual Meeting of the Ceramic Society of Japan (1985) 373.

49) Y. Ozaki, M. Wadasako and Y. Masumori: "Synthesis of ultrafine Sb_2O_3-Sb_2O_5 composite particle from metal alkoxides," Proc. 60th Annual Meeting of the Ceramic Society of Japan (1985) 375.

50) Y. Ozaki, M. Wadasako and J. Miyakoshi: "Synthesis of $(Ba, Sr)Nb_2O_6$ from metal alkoxides," Proc. 60th Annual Meeting of the Ceramic Society of Japan (1985) 377.

51) Y. Ozaki, M. Wadasako and T. Kasai: "Production of cramic microspheres from metal alkoxides," Proc. 60th Annual Meeting of the Ceramic Society of Japan (1985) 379.

52) Y. Ozaki, M. Wadasako and K. Hayashi: "Synthesis of low temperature glass powders from metal alkoxides (I)," Proc. 60th Annual Meeting of the Ceramic Society of Japan (1985) 381.

53) Y. Ozaki, K. Hayashi and Y. Imanishi: "Synthesis of low teperature glass powders from metal alkoxides (II)," Proc. 60th Annual Meeting of the Ceramic Society of Japan (1985) 383.

54) Y. Ozaki: "Macrostructure design of ceramics by metal alkoxides," Proc. 60th Annual Meeting of the Ceramic Society of Japan (1985) 385.

55) Y. Ozaki, Y. Akutsu and T. Tsurumi: "Preparation of ZnO varistor powder from matal alkoxides," Proc. 61st Annual Meeting of the Ceramic Society of Japan (1986) 145.

56) Y. Ozaki, Y. Akutsu, T. Kasai and S. Yamamoto: "Fabrication $BaTiO_3$ micro spheres from metal alkoxide," Proc. 61st Annual Meeting of the Ceramic Society of Japan (1986) 147.

57) Y. Akutsu and Y. Ozaki: "Preparation of Ag fine powders," Proc. 61st Annual Meeting of the Ceramic Society of Japan (1986) 149.

58) Y. Ozaki, Y. Akutsu and K. Kojima: "Synthesis of $(Pb, Ba, Sr)Nb_2O_6$ powder from metal alkoxides," Proc. 61st Annual Meeting of the Ceramic Society of Japan (1986) 151.

59) Y. Ozaki, Y. Akutsu and N. Kosugi: "Synthesis of RuO_2 powders for condctive paste from metal alkoxide," Proc. 61st Annual Meeting of the Ceramic Society of Japan (1986) 153.

60) Y. Ozaki, Y. Akutsu and T. Miura: "Synthesis of low melting glass powders from metal alkoxides," Proc. 61st Annual Meeting of the Ceramic Society of Japan (1986) 155.

61) Y. Ozaki, Y. Akutsu and T. Kasai: "Synthesis of magnetic fluid from metal alkuxides," Proc. 61st Annual Meeting of the Ceramic Society of Japan (1986) 157.

62) Y. Ozaki, Y. Akutsu and S. Yamamoto: "Fabrication of $BaTiO_3$ thin film from metal alkoxides," Proc. 61st Annual Meeting of the Ceramic Society of Japan (1986) 159.

63) Y. Ozaki, Y. Akutsu and K. Kawasaki: "Production of stabilized zirconia sphere from metal alkoxides," Proc. 61st Annual Meeting of the Ceramic Society of Japan (1986) 303.

10. Alumina Ceramic Substrates

Tamotsu UEYAMA* and Hiroshi WADA*

Abstract

This Paper describes the doctor blade method of manufacturing alumina ceramic substrates as well as their applications. Considerable emphasis is placed on finding optimum substrate conditions and techniques rendering those substrate conditions, to meet the requirements of electronics and computer industries. To meet these requirements, stricter control is urged, and successes to date, and some promising data are reported.

Keywords: alumina substrate, dispersion, casting, surface roughness, alumina powder, shrinkage.

10.1 Introduction

The largest market of alumina ceramics for electronic parts is substrates for ICs (integrated circuits) and LSIs (large scale integrated circuits). Rapid progress in increasing the density of these ICs has accompanied advancements in electronic parts manufacturing techniques. Recently VLSIs (very large scale integrated circuits) with over one million transistors mounted on a single silicon chip are mass produced and are broadly applied[1] to scientific calculators, cameras, watches and automotive vehicles besides electronic products for industrial use such as computers and computer controlled electronic switching exchangers for telephones. As for these substrates for locating ICs, plastics and glass plates are used for simple ICs for scientific calculators, cameras, etc.[2] However, these materials have not good thermal conductivity.[3] For computer controlled electronic switching exchangers for telepones and other electronic products for industrial use which are required to provide complicated and multiple functions, high density LSIs with complicated ICs on a single silicon chip are used. These LSIs are located on alumina substrates providing high thermal conductivity. Furthermore, in large size computers VLSIs with over one million transistors on a single silicon chip with a side length of 5 mm are used.[4-5] A number of these VLSIs are mounted on a substrate for use. With VLSIs of such super high density, power consumption is over 20 W/cm². And the calorific power is of a level is equivalent to that of an electric heater for home use. However, currently produced ICs, LSIs and VLSIs do not normally operate at over 125 °C. Accordingly, the substrates used for electronic parts should provide high thermal resistance, high thermal conductivity and superior mechanical and electrical characteristics in addition to high dimensional accuracy required for packag-

---

* Ibaraki Research Laboratory, Hitachi Chemical Co., Ltd., 1380-1, Aza-Nishihara, Oaza-Tarasaki, Katsuta, Ibaraki 312, Japan.

ing at high density.

Alumina ceramic substrates used for electronic circuits in general include substrates for thick films with alumina content of 96 %, non-glazed substrates for thin films with alumina content of 99.5 % and upwards, and glazed substrates with glass during application. The majority of substrates are produced by pressing or sheet casting. Sheet casting is divided into extruder method, calender roll method and doctor blade method. The doctor blade method is the major method used in casting because it easily permits increases in sheet size and high precision in sheet thickness, and provides superior dimensional stability.

The sheet casting method was developed as a method effective for the production of electronic ceramic materials. The sheet casting method was first applied to production of ferroelectric ceramics of TiO_2. It was reported a slip was cast on a moving belt in 1947. Since then, inroads were made in the development of alumina green sheets for IC substrates mainly in the U.S.A., by J. L. Park, J. J. Thompson, H. W. Stetson, W. J. Gyurk, etc. in the 1960's.

Manufacturing of alumina ceramic substrates by the doctor blade method and their applications are described in this section. The doctor blade method is most suitable and effective for production of alumina green sheet.

10.2 Manufacturing Method

An outline of the processes occurring in the manufacture of alumina ceramic substrates by the doctor blade method is shown in Fig. 10.1. The average particle size of

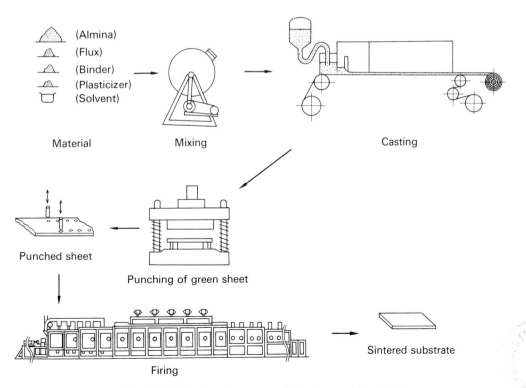

Fig. 10.1 Manufacturing process of alumina ceramic substrate.

alumina powder used to create substrates for thick films is in many cases 1-3 μm in diameter. The particle size, particle size distribution, bulk density, specific surface area, etc. of the alumina powder and of flux (sintering promotor agent) directly effect the final product's quality. Accordingly, alumina powder and flux are severely inspected and adjusted. Slip is produced by adding binder, plasticizer, solvent and deflocculant to these materials. Green sheets are produced by spreading this slip on polyester films or sheets of paper coated with surface lubricant, whereupon they are punched with dyes to produce arbitrary forms and then fired to produce ceramic substrates. The details of this manufacturing method are described next.

10.2.1 Materials
(a) Alumina Powder

Alumina powder produced by the Bayer method is most frequently used as material for alumina ceramic substrates. This is because alumina powder produced by the Bayer method is most stable in both quality and price. Many primary particles aggregate and secondary particles are formed as shown in Fig. 10.2. In this agglomerated condition, alumina powder influences sinterability as well as influencing the characteristics of the sintered substrates.[6] The characteristics of materials and ceramics required for controlling the quality of ceramics are shown in Table 10.1. The sinterability and electrical characteristics of ceramics are largely affected by the chemical composition of the materials and also

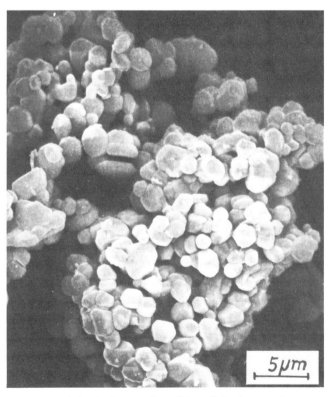

Fig. 10.2 Agglomerated condition of alumina powder.

by impurities in the materials.[8] The alkali component of impurities makes insulation resistance and tanδ inferior. Evaluations made during the formation of thick films, show that scattering of the resistance value increases with increases in the calcia content.[9] When evaluation is made on particles size, particle size distribution, specific surface area and so forth, the sintering shrinkage ratio appears dependent on the particle size, but even if the averaged particle size is the same, the shrinkage ratio varies if the particle size distribution changes. Furthermore, surface roughness is larger when the particle size is larger. In addition, the green density and bulk density of sintered ceramics are correlated with the total surface area of the powder, and the ratio of the specific surface area to the average particle size obtained through measurement of particle size distribution is a criterion for the agglomerated energy of the powder.[10] Thus, particle size, particle size distribution and specific surface area of the powder have significant effects. The agglomerated energy and packing density of the powder can be evaluated from bulk density.[6] The result of this evaluation is closely related to the extent of agglomeration, green density and bulk density of sintered ceramics.

Table 10.1 Characterizations of alumina ceramics

Characterization of Alumina Powders	Characterization of Green Sheets	Characterization of Sintered Bodies	Evaluation Item of Substrate
Purity	Pucking density	Composition	Thermal properties
Packing density	Uniformity	Bulk density	Mechanical properties
Specific Surface Area	Unisotropy	Porosity ratio	Electrical properties
Internal Stress	Internal Stress	Distribution of porosity	Optical properties
Unifomity	Strength	Porosity size	Chemical properties
Particle Diameter	Flux composition	Grain size	Physical properties
Distribution of Particle Diameter.			
Agglomerated Particle Diameter.			

The characteristics of sintered bodies are largely affected by the characteristics of the powder as explained above. It is therefore important to sufficiently evaluate and control the powder characteristics in order to improve and stabilize the quality of sintered bodies.[7] These powder characteristics can be evaluated by the following methods.

(1) Chemical composition, impurities and crystal structure: X-ray diffraction, X-ray flourescent analysis, atomic absorption analysis, etc.

(2) Particle form, size, size distribution and specific surface area: Observation by microscope, various measurements for particle size distribution, various measurement for specific surface areas.

(3) Agglomerated energy and packing property: Various measurements of powder density, contact angle, sedimentation volume, etc., powder tester, and so forth.

(b) Flux (Sintering Promotor Agents)

A high temperature of 1,700–1,750 °C is required for densely sintering when alumina particle size 1–3 μm in diameter. This required temperature level is higher than the maximum temperature 1,500–1,650 °C of industrially mass produced furnaces. For usual alumina ceramics, therefore, other substances are added in amounts that do not have a major adverse affect on the characteristics of alumina, as alumina's flux, to accelerate liquid phase sintering in order to reduce the sintering temperature. However, the type and composition of the flux exert great influence on the characteristics of substrates for thick films, and therefore, their selection is also very important.[16] In concrete, it is necessary to bond conductors and resistors firmly on the substrates and scattering of resistance values should be suppressed. The thick film paste contains boron silicate glass in addition to metals and metal oxides, to increase the bonding force. The glass in the paste causes reaction at low temperatures (800–1,000 °C) with the glassy phase and secondary crystal phase produced as a result of reaction between alumina and flux located on the substrate surfaces and bonded firmly to the substrates.[13] When high-purity alumina, contains little flux, is used as substrate, therefore, thick film conductors and resistors have almost no bonding force on it. The flux generally used for this bonding effect are silica, magnesia and calcia. All of these substances lower the forming temperature of liquid phase. Magnesia in particular suppresses grain growth of alumina. Calcium oxide, on the other hand, increases the scattering of resistance values of thick film, and therefore, amounts added should be limited. The sinterability of alumina ceramic substrates is also affected by the properties of the flux materials, that is, the properties of silica, basic magnesium carbonate, magnesium oxide, calcium carbonate, talc, dolomite, etc. besides the proportion ratio of the flux.

(c) Binder and Auxiliary Materials

The functions of the binder and of the plasticizer are important in manufacturing flexible sheets by the doctor blade method and in producing products of high dimensional accuracy. Numerous binders, plasticizers, etc., have been proposed since ferroelectric ceramics of TiO_2 were manufactured by the doctor blade method in 1947. Commonly used binders, plasticizers and dispersants are shown in Table 10. 2.[15] The characteristics required of binders are as follows.[12]

Table 10. 2 Commonly used binders, plasticizers and dispersants.

Binders	Plasticizers	Dispersants	Soluvent
	(S·A·I·B)		
Cellulose acetate	Sucrose acetate isobutyrate	Menhaden fish oil	Trichlore Ethylene
Polyacrylate	Glycerin	Octadecylamine	Alcohol
Polymethacrylate	Dibutyle phthalate	Glycerylmonooleate	Ethyle Acetete
Poly (vinyl alcohol)	Diisodecyl phthalate	Glyceryltrioleate	Toluen
Poly (vinyl butyral)			Aceton
Poly (vinyl acetate)	Polyethylene Glycole		Methyl Ethyl Ketone
			Water
Mlthylcellulose	Dioctyl phthalete		

(1) It should be able to add plasticity to sheets with an addition rate less than 8-10 %.
(2) Substrates and gases produced as a result of decomposition during sintering should be harmless.
(3) No fired residues should be produced.
(4) It should be possible to use non-flammable volatile solvent of a low price.
(5) It should provide weather resistance and sheets should not deteriorate during storage.
(6) The price should be low.

Any binder satisfying the characteristics indicated above is considered an ideal binder. The current binders used industrially in the doctor blade method are butyral resin, acrylic resin derivatives, polystyrene derivatives, cellulose derivatives, etc.[14] In addition, nonaqueous solvents are generally used. The reason for this is that dimensional stability of sheets is adversely affected by the moisture present in the air when aqueous solvents are used. Furthermore, due to the fact that the affinity of materials (powders) with water is large, the viscous characteristics of the slip deviate greatly from newtonian. Recently, however, with the large progress made in technologies and improvement of working environments, many manufacturers are re-examining the use of aqueous solvents with no pollution and safety as the objectives.

10.2.2 Mixing of Slury

The mixing process is that used to manufacture a homogeneous mixture of ceramic powder, binder, plasticizer, solvent, dispersant and so forth. It is the most important process because conditions of pulvarization, mixing, dispersion, degassing and so forth affect each other in a very complicated way.

The current mixers used for manufacturing slip are ball mills, sand grinders, affrition mill, etc. Among these, ball mills are most widely used. In the ball mill mixing, mills and balls of high alumina content are used to prevent the entry of various powdered substances produced as a result of wear and abrasion during mixing. Differences in the mixing conditions causes scattering of sheet thickness, shrinkage stability at the time of sintering, scattering of density and so forth.

A mixture of various solvents is usually selected in order to thoroughly dissolve the binder and plasticizer and to prevent problems which would otherwise occur during drying.

Because organic solvents are used it is necessary that all the equipment used be explosion-proof, to provide a high degree of safety. Solvent recovery systems are also used.

As the slip in which ceramic powder is homogeneously dispersed contains numerous air bubbles which enter during the mixing process, it is necessary to carry out degassing. While degassing is performed with agitating the slip, the extent of the degassing should be considered carefully because the solvent formation changes during the degassing operation. Examples of changes in solvent formation are shown in Fig. 10.3. If the solvent formation changes, the evaporation velocity at the time of drying varies, and troubles such as cracking in the green sheets may occur in the drying process. In addition to eliminating air bubbles in the slip, the purpose of degassing is to adjust the viscosity of the slip for the next process.

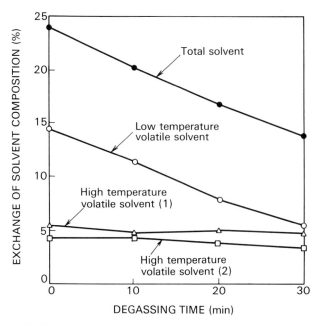

Fig. 10.3 Degassing condition and composition exchange of solvent.

10.2.3 Casting

The usual sheet manufacturing processes are shown in Fig. 10. 4. The slip, with its viscosity adjusted in the degassing process, is filtered through a strainer and fed to a sheet manufacturing machine, usually called a casting machine, where the slip is continuously converted into sheets. As the quality of substrates, including dimension precision warp and so forth, is almost entirely determined at the time when sheets are cast, this process also very important. Sheet thickness is determined in this process, while sintering shrinkage ratio, uniformity and density of products are determined. The factors for controlling thickness include (1) slip viscosity, (2) casting speed, (3) doctor blade gap, and (4) slip head height. The factors which determine shrinkage ratio and density are (1) temperature profile in the drying zone and (2) the speed at which solvent is vaporized and so foth.

The slip produced in the preceding process is applied to a specified thickness on a polyester film. Then the solvent is gradually caused to vaporize, and the residual dried green sheet is peeled off from the polyester film and coiled after being cut into specified widths. The thickness of the green sheet is determined by the doctor blade gap and the film speed. In general, the speed is determined by the drying capacity of the equipment, and the thickness is determined by the gap.

Alumina green sheets are controlled by the following qualities in the intermediate process.

(1) Sintering shrinkage and anisotropy
(2) Uniformity of thickness
(3) Smoothness of surface

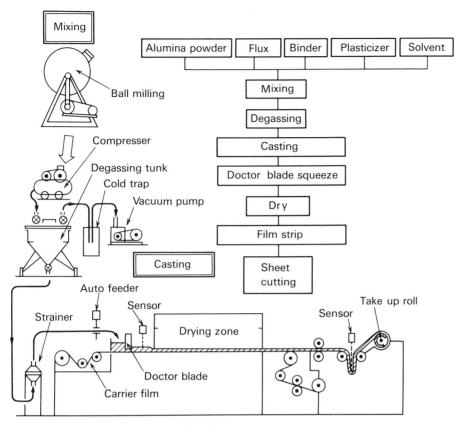

Fig. 10.4 Schematic of the manufacturing process of green sheet by doctor blade method.

(4) Tensile strength
(5) Bonding strength
(6) Compression property
(7) Gas permeability
(8) Property of de-binder

Much stricter control over the characteristics indicated above is required these days, to meet the increasing demand for substrates for use as electronics materials requiring a guaranteed high precision of dimension for multilayered substrate[17] and sintered products. A very important subject at the present time is how to accomplish such strict control and at the same time keep costs low. The relationship between the green density and shrinkage is shown in Fig. 10.5. Here, changes in the density before and after sintering coincide with the sintering shrinkage ratios in the three-dimensional direction. The density after sintering is controlled to a high level of accuracy by sintering conditions. Therefore, it can be said that if the density of green sheets is equalized to a high level of accuracy, it is possible to stabilize the shrinkage ratio. However, anisotropy is provided in the shrinkage in the three-dimensional direction depending on the powder conditions, and there are cases where this anisotropy varies according to the green sheet thickness. In order to manufacture sintered products with high dimensional accuracy in all three dimensions, i.e., length, width and depth, it is necessary to control the three-dimensional anisotropy, so that it is suitably balanced, including even shrinkage in the direction of thickness besides the density of the green sheets. In general, dimensional accuracy is ± 1 %. However, accuracy of ± 0.5 % has already been provided, and there are reports which indicate that 0.1 % was accomplished for multilayer substrates for computers. Thus, the technology for stabilizing the shrinkage ratio is rapidly progressing.

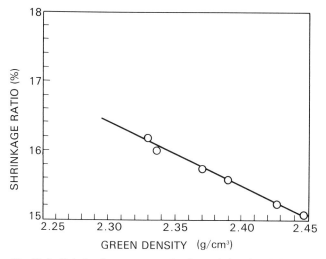

Fig. 10. 5 Relation between green density and sintering shrinkage.

10. 2. 4 Sintering

Most important in sintering is how to efficiently and quickly eliminate organic materials in the green sheets regardless of the packing conditions of the inorganic materials. When organic polymer is heated to high temperature, it is decomposed and a part or the whole of its components are generally lost. If it is heated in the air, a part of the polymer is oxidized by the oxygen present in the air, and oxides such as hydroperoxide and carbonyl group are produced.[18] Accordingly, macromolecules are decomposed and the molecular weight drops. When the temperature becomes close to the boiling points of oxidation products, they vaporize and the weight drops. On the other hand, no oxidation occurs if heating is made in an inert gas atmosphere where no oxygen is present, but when the temperature rises the macromolecules are subject to thermal decomposition, and breakage of the main chain, branches and crosslinkage occurs, causing the macromolecules to change gradually into low molecular weight compounds and vaporize. It is important to check the thermal decomposition curves of the organic materials used with DTA and TGA, and determine the temperature profiles of de-organic materials which match with these curves.

A typical continuous sintering furnace is shown in Fig. 10. 6. Products are sintered in the individual state or in the state where multiple products are stacked with sintering efficiency and forms taken into account. Sintering made after removing organic materials to a certain extent in advance, is effective to increase sintering efficiency. The thermal

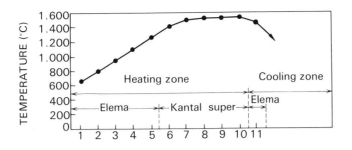

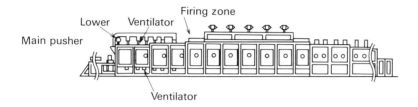

Fig. 10. 6 Temperature profile and outline of pusher type belt furnance.

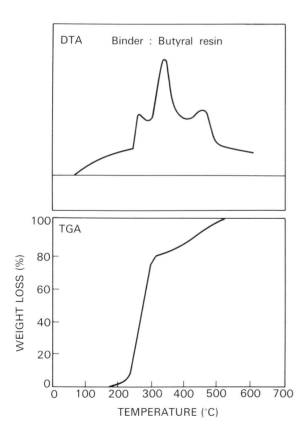

Fig. 10. 7 DTA and TGA characteristic of green sheet.

decomposition curve, thermal expansion and bending strength of the green sheet produced using a representative butyral resin, are shown in Figs. 10. 7, 10. 8 and 10. 9, respectively. As these figures show, green sheet shrinks rapidly and its strength drops extremely when the binder and plasticizer contained in the sheet begin to vaporize and decompose. It is therefore necessary to carry out sintering slowly in the temperature rage where conditions of the sheet change most rapidly. If the temperature rising rate is too large, the cracks shown in Fig. 10. 10 will occur as the resin is removed from the external circumference as shown in Fig. 10. 11. In the sintering of alumina ceramics, if sintering is made in the state where de-binder operation of the green sheet is insufficient, carbon will remain in the ceramics as a decomposition residue of the binder. This carbon reacts with Al_2O_3 to produce Al-O, Al-O-C in the temperature region of 1,250–1,650 °C, which is the alumina green sheet

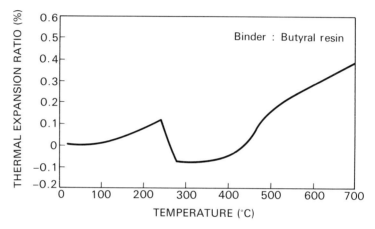

Fig. 10. 8 Thermal expansion curve of green sheet.

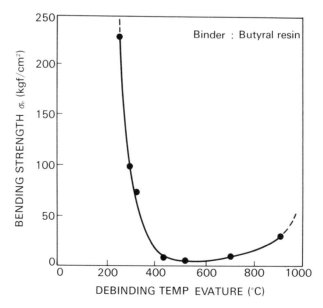

Fig. 10. 9 Debinding temperature and bending strength of green sheet.

sintering temperature, and voids are produced in the sintered bodies due to evaporation and decomposition of these substances.[13] In order to obtain sintered bodies of high density, therefore, it is necessary to minimize formation of these compounds.

Fig. 10.10 Critical crack of sintered substrate.

Fig. 10.11 Debinder condition of green sheet.

10.3 Features of Ceramic Substrates

10.3.1 Surface Roughness

SEM pictures and surface roughness data of an alumina ceramic substrate are shown in Fig. 10.12. It is desirable that the surfaces of ceramic substrates for thick films are smooth in order to eliminate bleeding or choking at the time conductors and resistors are printed. However, if these surfaces are extremely smooth, the ink wetting is inferior, and therefore, ceramic substrates of surface roughness 1.5 to 3.5 μm are usually used. For ceramic substrates for thin films, on the other hand, surface roughness less than 0.2 μmRa is preferable because disconnection tends to occur unless the surfaces are smooth, due to the extreme thinness of evaporation films.

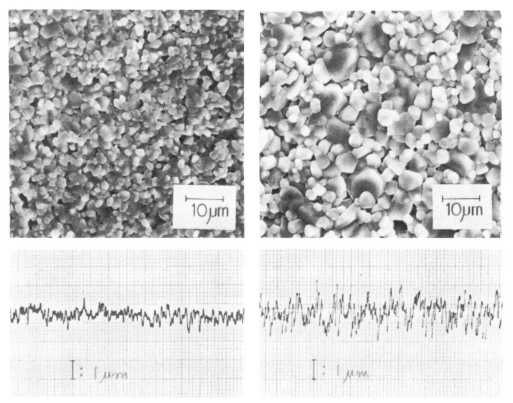

Fig. 10.12 Surface conditions of the substrate.

The surface roughness of substrates is proportional to the crystal size of the ceramic powders. For production of alumina ceramic substrates, therefore, the smaller the particle size of alumina powder the better. In general, however, alumina powder consists of tens to hundreds of particles solidly agglomerated to each other, and this agglomeration is stronger when the particle size is less.

10.3.2 Precision of Dimensions

Alumina ceramic substrates for electronic parts are required to provide high precision dimensions, and strict control is required over their production. The big problem is how to manufacture them at low costs. With ceramics the changes in the voids before and after sintering coincide with changes in volume, but they are three-dimensional changes, and ceramic substrates are required to have constant shrinkage in length, width and thickness, overall. It is therefore necessary to ensure that three-dimensional anisotropy is suitably balanced with the substrate materials and manufacturing processes sufficiently taken into account. The precision of dimensions is usually ± 1 %, however a precision of ± 0.5 % has already been accomplished, and it was reported ± 0.3-± 0.05 % has already been accomplished with multilayer ceramic substrates for computers. Thus, shrinkage ratio stabilizing technologies have progressed greatly.

10.3.3 Others

Alumina ceramic substrates are required to provide various characteristics corresponding to their different applications. In evaluating them by formation of thick films for ceramic substrates for thick films, there are cases where the scattering of resistance values becomes large or the bonding strength drops due to reactions between glass components in the resistors and the glassy phase or secondary crystal phase located on the substrate surfaces or grain boundaries. Furthermore, voids on the surfaces and in the interior are problems for high density through hole substrates, ceramic substrates for thermal printer heads, glazed ceramic substrates and so forth.

10.4 Applied Products

Alumina ceramic substrates are used in large quantities for thick film hybrid ICs and multilayer ceramic packages for electronics products such as VTRs, TVs and tape recorders, cameras and electronic devices for automotive vehicles. Because alumina ceramic substrates provide advantages such as (1) high insulation resistance and the non-occurance of moisture absorption deterioration, (2) good thermal diffusibility of ICs due to superior thermal conductivity, and (3) wiring density is easily increased and multiple functions are possible, both the volume of use and applications are on an upward trend which promises to continue well into the future. Furthermore, alumina ceramic substrates of surface roughness 0.2 μmRa or less are expected to be used increasingly in the fields of thin film substrates for office automation equipment and for space communication. The thermal printer recording electrode head shown in Fig. 10.13 is an example of such an alumina ceramic substrate. It is used with glaze glass applied to the surface of a high density alumina ceramic substrate. The chip carrier shown in Fig. 10.14 is a kind of package produced for mounting naked semiconductor devices. It permits standardization of ICs, and in addition, it contributes to the size reduction of composite packages as well as increasing the performance and reliability of substrates.

A multilayer ceramic substrate is a substrate produced by laminating and unilizing green sheets with conductor patterns printed on them and by co-firing them. Multilayer

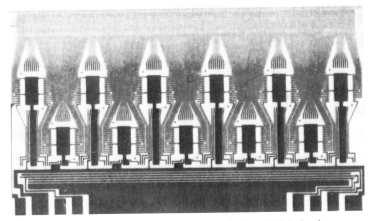

Fig. 10.13 Thermal printer head for facsimil used high density alumina substrate.

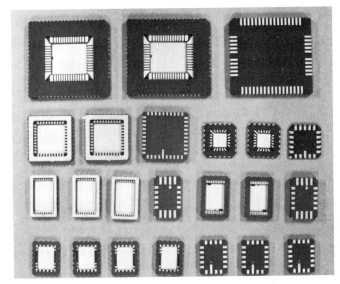

Fig. 10.14 Chip carrier for semiconductor device.

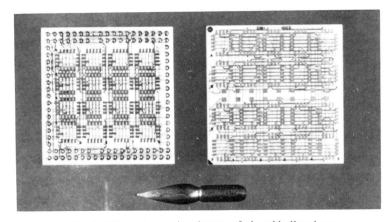

Fig. 10.15 Multilayered ceramic substrate of pin grid allayed type for large size computers use.

ceramic substrates of pin grid allayed type, used in large-size computers, are shown in Fig. 10. 15. They are 8-layer substrates with a fine line of 0.1 mm and with a spacing of 0.1 mm and are connected at through holes of 0.1 mm in diameter. Providing wiring in extra density is easy with substrates produced by this method, and the resulting permit the direct packaging of many LSIs and VLSIs. Accordingly, the demand for substrates of this type is rapidly increasing.

10. 5 Conclusion

Alumina ceramics for electronics materials are ceramics which provide superior multiplicity and which are capable of producing various properties when adapted to various applications by changing the chemical composition of the materials and also by changing their forming processes through the sintering processes described earlier. Accordingly, fine alumina ceramics are ceramic materials most suitable to meet the strict requirements of high performance and high precision necessitated by the increase in density and reliability of electronic products and electronization of mechanical industries. Because of, and in view of the nature of the demand, therefore, development of multiple and diversified applications will further make advance and expand further in the future.

Once the relationship between powder characteristics and sintering conditions, and microstructure and characteristics of sintered bodies is completely understood, improvements in characteristics will be accomplished and it will become possible to develop novel applications which cannot be considered at the present time.

References

1) A. K. Malhatra et al.: "Fine start, a new concept in VLSI package," Hewlett Packard. J., **34**, 8 (1983) 24-26.
2) W. O. Geesfeldt: "The application of thick film technology in multilayer circutry," Proc. Int. Electronic Circuit Packaging Symp. (1967).
3) Yamagishi: "IMST substrate and porcelained substrate," Electronics Materials, 5 (1980) 79 [in Japanese].
4) P. N. Venkatachalam: "Pulse propagation properties of multilayer ceramic multichip module for VLSI circuits," Electronic Components Conf. (1983) 33.
   B. Schwartz: "Micro electronics packaging, II," J. Am. Ceram. Soc. Bull., **63**, 4 (1984) 577-581.
5) A. Blodgett, Jr.: "A multilayer ceramic multichip module," IEEE Trans. Components, Hybrids & Manuf. Technol., **CHMT-3**, 4 (1980) 634-637.
6) H. P. Cahoon et al.: "Sintering and grain growth of alpha-alumina," J. Am. Ceram. Soc., **39**, 10 (1956) 337-344.
   L. Berrin et al.: "High purity reactive alumina powders: I, chemical and powder density," J. Am. Ceram. Soc. Bull., **51**, 11 (1972) 840-844.
7) "Fundamentals of ceramics engineering," Ceramics Jpn., **16**, 4 (1981) 296 [in Japanese].
8) R. A. Vermetti et al.: "Effect of metal oxide additions on the high-temperature electrical conductivity of alumina," J. Am. Ceram. Soc., **49**, 4 (1966) 194-199.
9) J. R. Larry et al.: "Thick Film Technology, An Introduction to the Materials," IEEE Trans. Components, Hybrids & Manuf. Technol., **CHMT-3**, 2 (1980) 211-225.
10) D. W. Johnson et al.: "High purity reactive alumina powders: II, particle size and agglomeration study," J. Am. Ceram. Soc. Bull., **51**, 12 (1972) 896-901.
11) H. D. Rigterink: Electronic Ceramics, The

American Ceramics Society (1969).
12) J. J. Thompson : "Forming thin ceramics," J. Am. Ceram. Soc. Bull., **42**, 9 (1963) 480–481.
13) F. J. Klug et al. : "Microstructure development of aluminum oxide; graphite mixture during carbothermic reduction," J. Am. Ceram. Soc., **65**, 12 (1982) 619–624.
14) K. Ettre and G. R. Castles : "Pressure-fusible tapes for multilayer structures," J. Am. Ceram. Soc. Bull., **51**, 5 (1972) 482.
15) D. J. Shanefield and R. E. Mistler : "Fine grained alumina substrates: I, the manufacturing process," J. Am. Ceram. Soc. Bull., **53**, 5 (1974) 416.
16) R. C. Buchanan : "Substrate surfaces for microelectronics," Ceramic Age (April 1968) 60.
17) B. Schwartz and D. L. Wilcox : "Laminated Ceramics," Ceramic Age (June 1967) 40.
18) R. B. Mesrobian and A. V. Tobolsky : "Some structural and chemical aspects of aging and degradation of vinyl and diene polymers," J. Poly Sci., **2**, 3 (1947) 463.

11. Bonding between Ceramics and Metals by HIPing

Masahiko SHIMADA[*1], Katshuaki SUGANUMA[*2], Taira OKAMOTO[*2] and Mitsue KOIZUMI[*2]

Abstract

Internal stress due to thermal expansion mismatch between alumina and steel with and without an interlayer is evaluated here. On the basis of the results of calculation, solid-state bonding experiments with alumina and steel using the interlayer method are carried out at 1,000–1,400°C under a pressure of 100–200 MPa for 30 min using the hot isostatic pressing technique. When a laminated interlayer (niobium/molybdenum) is used, a joint with high bonding strength is obtained. Results of thermal cycle tests at temperatures from 25°C to 500°C reveal that the laminated interlayer is superior as an interlayer material. From these results it is felt that strong bonding has been accomplished, and that the HIPing technique is a useful method for diffusion bonding of ceramic-metal.

Keywords: solid-state bonding, ceramic-metal joint, HIPing.

11.1 Introduction

There has been considerable interest recently in the use of ceramics such as Si_3N_4, SiC and PSZ for engineering purposes. Research and development have concentrated on the preparation of starting powders and fabrication of ceramic bodies.

The technique of bonding between ceramic and metal components is important in producing high-temperature structural materials with complicated shapes. There are many bonding methods, e.g. liquid phase bonding, eutectic bonding, thermocomponents.[1] Recently, a hot isostatic pressing technique (HIPing) has been developed for bonding ceramic-ceramic and ceramic-metal components.[2] For high temperature applications, it is necessary that bonded materials have good termal shock resistance. Since ceramics and metals usually have different coefficients of thermal expansion, internal stress acts across the interface between bonded materials.[3] Resistance of materials to thermal shock is one of the most important factors in determining reliability of bonded materials.

There are many reports on studies of bonding ceramic-ceramic and ceramic-metal by adding other components between bonded materials as liquid phase bonding, but few on

---

[*1] Department of Applied Chemistry, Faculty of Engineering, Tohoku University, Aza Aoba, Aramaki, Sendai, Miyagi 980, Japan.
[*2] Institute of Scientific and Industrial Research, Osaka Uiversity, 8-1, Mihogaoka, Ibaraki, Osaka 567, Japan.

diffusion bonding. In developing a method for producing strong bonded material with sufficient thermal shock resistance,[4,5] calculation of internal stress due to thermal expansion mismatch within a joint area has shown a new method of "laminated interlayer bonding," in which the interlayer is a composite of the materials to be bonded, in other words two kinds of thin metals. The chief aims of the present paper are to evaluate stress concentration in bonded interface due to thermal expansion mismatch and to enumerate the advantages of interlayer bonding methods. Alumina and a type of ferritic steel were selected as materials to be bonded in the present study.

11.2 Experimental Procedure

Computer calculation of internal stress within a joint area was conducted by DISAP-6, a finite element program put together by NEC; the model was a cylindrical butt-joint type. The element was the rectangular type. Figure 11.1 shows a joint model consisting of a single interlayer. Table 11.1 summarizes physical constants used in computer calculation. All physical constants used in the present calculation were assumed to be independent of temperature. Four kinds of interlayers, three pure metals and one cermet, were selected. The temperature difference (ΔT) was 1,000°C. Calculations were performed

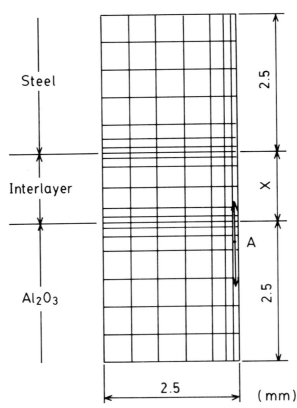

Fig. 11.1 Joint model consisting of single interlayer X for finite element method. (A: region of the maximum tensilestress)
[after K. Suganuma et al.[5]]

Table 11.1 Values of young's modulus (E), shear modulus (G), Poisson's ratio (ν) and thermal expansion coefficient (α). [after K. Suganuma et al.[5]]

	E (MPa)	G (MPa)	ν	$\alpha(\times 10^{-6})$
Al_2O_3	327,000	147,000	0.27	8.1
Steel	196,000	78,400	0.25	14.0
Nb	98,000	36,300	0.37	8.1
Ti	108,000	40,200	0.34	11.0
Mo	323,000	118,000	0.38	5.7
Fe-Al_2O_3	284,000	113,000	0.26	11.0

under following assumptions: (1) Materials behave elastically and there is no plastic deformation. (2) Physical constants of Fe-50 vol% Al_2O_3 are calculated using the linear law of mixture.

After polishing the surfaces of the ceramics and metals with several grades of diamond paste (8-0.3 μm), solid-state bonding experiments were conducted using the hot isostatic pressing technique. The two types of bonding carried out are illustrated in Fig. 11. 2: Metal/ceramic structures (A) for metallographic observations, and metal/ceramic/metal structures (B) for tensile and bending testing. For HIP bonding, the starting samples were put into a BN capsule, which was put into a container made of pyrex glass, as shown in Fig. 11. 3. The glass container was evacuated to 0.1 Pa and sealed. The glass sealed speciman was placed in the hot zone of a high-pressure vessel, and HIP bonding experiments were performed using Ar gas. The specimen was heated to the softening point of glass under a pressure of 2 MPa; then pressure and temperature were simultaneously raised to the 100-200 MPa and 1,000-1,400°C as illustrated in Fig. 11. 4. After the specimen was exposed to the desired experimental conditions, the container was stripped and the specimen cleaned by polishing the surface.

After the bonded samples were cut perpendicular to the bonding interface with a

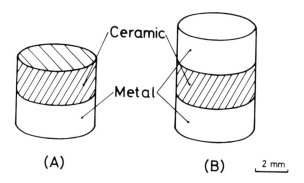

Fig. 11. 2 Schematic illustrations of two type of bonded samples. (A) for the observation of microstructure, and (B) for the strength test.

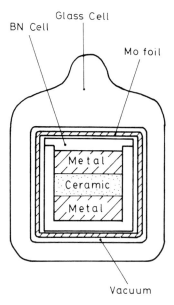

Fig. 11.3 The glass sealed sample for HIP bonding experiment.

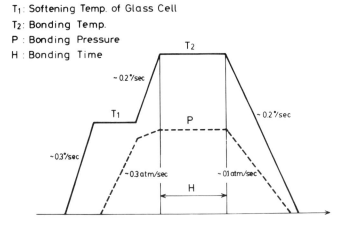

Fig. 11.4 Diagram of HIP bonding process.

diamond cutter, their microstructure was observed by scanning electron microscopy (SEM).

The strength of the bonded specimens was evaluated in terms of a tensile test and a bending test. 5 mm diameter and 6 mm height joints were used for the tensile test after the two threaded shoulders were bonded each end of the joints. The cross head speed was 0.5 mm/min. Bending specimens were cut from 7 mm diameter and 15 mm height joint; their dimensions were $2 \times 2 \times 15$ mm. As for the measurement of bending strength, one side of the sample surfaces on which the highest tensile stress introduced by bending test was polished into optical flatness. In the present bending strength experiments, a 4-point bending test was adopted with a cross head speed of 0.5 mm/min.

Resistance to thermal cycling was examined by a 4-point bending test using the samples subjected to thermal cycle. The bending specimens were sealed into silica tube in a vacuum, below 10^{-5} Torr. The tubes were put in and pulled out of a furnace controlled at 500°C isothermally. Heating and cooling rates were about 1 deg/sec and the keeping time was 2 min. The duration of one thermal cycle was about 20 min.

11.3 Results and Discussion

When interlayers were not used, the maximum tensile stress in alumina reached about 940 MPa in the region of element A, shown in Fig. 11.1. As tensile strength of the alumina was about 400 MPa, calculated tensile stress of 940 MPa was too high; bonded material failure is ascribed to internal stress. Figure 11.5 shows the effects of thickness of four kinds of interlayers on maximum tensile stress. It is not that niobium is the most effective interlayer material for reducing internal stress as it has the same coefficient of thermal expansion as that of alumina. Molybdenum, which has a smaller coefficient of thermal expansion than that of alumina, and titanium and the cermet, whose coefficients of thermal expansion are larger than that of alumina, were also effective in reducing the maximum tensile stress, as shown in Fig. 11.5. The present bonding experiment between alumina and steel was carried out using a 1 mm niobium interlayer.

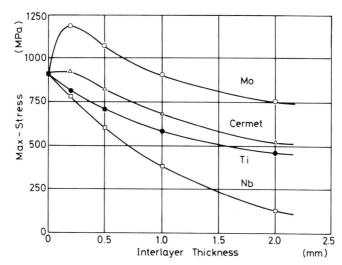

Fig. 11.5 Interlayer thickness dependence of the maximum tensile stress. [after K. Suganuma et al.[5]]

Calculated internal stress was about 400 MPa. Mean tensile strength of three samples was 5.6 MPa. Unfortunately, the measured value was relatively low because a great deal of internal stress still remained. This fact indicates the need to modify the type of interlayer in order to make a high strength bond. For this purpose one more layer of which the thermal expansion coefficient is smaller than those of niobium and alumina is inserted between niobium and steel. Molybdenum was selected for this layer. Maximum tensile stress in the joint area using a laminated interlayer, of which the total thickness was 1 mm, was calculated and stress as a function of molybdenum thickness is shown in Fig. 11.6. It is noted that there is an optimum ratio of niobium to molybdenum thickness to keep internal stress to a minimum.

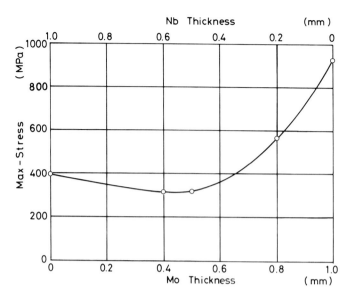

Fig. 11. 6 The maximum stress as a function of molybdenum thickness. [after K. Suganuma et al.[5)]]

A representative example of solid-state bonded sample of steel-Al_2O_3 system bonded at 100 MPa and 1,300°C for 1 h is shown in Fig. 11. 7. Steel and alumina cannot be bonded by the conventional diffusion method or even by high pressure bonding technique; insufficient reaction and a large difference in thermal expansion coefficients prevent this. On the basis of calculation results of internal stress due to thermal expansion mismatch between alumina and steel with and without an interlayer, bonding experiments between alumina and steel using interlayer method were carried out at 1,300°C and 100 MPa for 1 h. As seen in Fig. 11. 7 (A), when two kinds of interlayer materials such as Mo and Nb metals were used, bonding between alumina and steel was achieved.

Scanning electron micrographs of the interface of bonded sample is shown in Fig. 11. 7(B). It is generally said that conventional fusion welding often results in significant microstructural change at the interface zone. In the present results, as seen in Fig. 11. 7(B), no significant microstructural change near the interface was found.

Room temperature tensile and bending strength of joints with 1 mm niobium and with 1 mm niobium/molybdenum are listed in Table 11. 2. Results of the tensile strength indicated that the strength of the joint with 1 mm niobium as interlayer was very weak, but it was significantly high, 63 MPa, for the joint with a 1 mm niobium/molybdenum interlayer. The bonded sample with 1 mm niobium interlayer had a fracture path within alumina due to the residual stress, but the joint with 1 mm niobium/molybdenum interlayer was fractured along the alumina/niobium interface. Bending strength of the bonded sample was much higher than tensile strength. This difference is thought to be due to the effect of residual stress within the as-received joint, caused by thermal expansion mismatch. It is believed that, by cutting the joint, this stress is relieved. The value of 512 MPa in the joint with 1 mm niobium/molybdnum interlayer was almost as strong as that of alumina used in

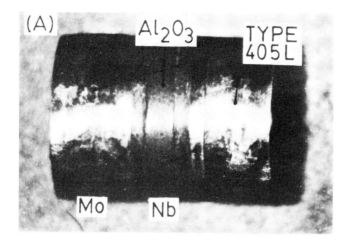

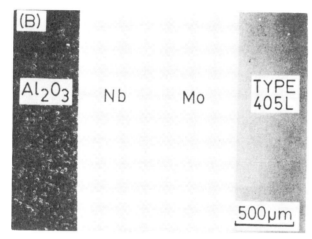

Fig. 11.7 Bonded sample of Al$_2$O$_3$-steel (A) and scanning electron micrograph of the interface (B).

Table 11.2 The tensile and 4-point bending strength of Al$_2$O$_3$-steel joint with interlayer.

Interlayer	Nb (1.0 mm)	Nb/Mo (0.5 mm + 0.5 mm)
Tensile Strength	5.6 MPa	63.0 MPa
4-point Bending Strength	357 MPa	512 MPa

the present study.

Figure 11.8 shows the change of bending strength due to thermal cycling between room temperature and 500°C. The strength degradation of the joint with 1 mm niobium was apparent even after one thermal cycle. Cracks in alumina were sometimes observed in the specimen subjected to thermal cycling. The joint with 1.5 mm niobium interlayer was superior, but a slight degradation was noticed after 100 cycles. On the other hand, the resistance to thermal cycling of the bonded sample with a niobium/molybdenum interlayer was excellent; there was no strength degradation.

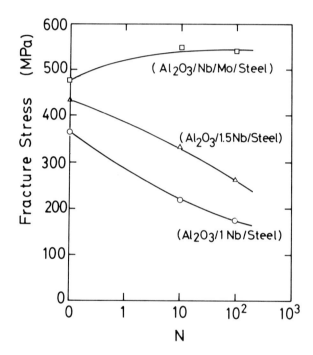

Fig. 11.8 Resistance to thermal cycling between room temperature and 500°C.

In the present study, internal stress due to thermal expansion mismatch between alumina and steel with and without an interlayer has been evaluated. The bondings using two kinds of interlayer materials, considered effective interlayers for bonding from the results of the calculation, were carried out. The model of these interlayers is generalized by taking account of thermal expansion coefficient. Effects of interlayers on bonding strength and resistance to thermal cycling for alumina/steel joints have been examined in the present work. The interlayers were 1 mm or 1.5 mm niobium and 1 mm niobium/molybdenum laminated layer. Results of strength test and the thermal cycling test indicate that the laminated interlayer is superior. Bending strength of joints with the laminated layer bonded at 1,400°C under a pressure of 100 MPa was the same as that of an alumina body; it resists thermal cycling at temperatures between room temperature and 500°C.

References

1) W. A. Owczarski and D. F. Paulonis: "Application of diffusion welding in the U.S.," Weld. J. (Miami), **60**, 2 (1981) 22–33.
2) H. D. Hanes: High-Pressure Science and Technology, Vol. 2., ed. by K. D. Timmerhaus and M. S. Barber, Plenum Publishing Co., New York (1979) 630–650.
3) T. Iseki and M. G. Nicholas: "The elevated temperature strength of alumina-aluminium and magnesium-aluminium samples," J. Mater. Sci., **14**, 3 (1979) 687–692.
4) K. Suganuma, T. Okamoto, M. Shimada and M. Koizumi: "New method for solid-state bonding between ceramics and metals," Commum. Am. Ceram. Soc., **66**, 7 (1983) C-117-C-118.
5) K. Suganuma, T. Okamoto, M. Shimada and M. Koizumi: "Effect of interlayers in ceramic-metal joints with thermal expansion mismatches," Commum. Am. Ceram. Soc., **67**, 12 (1984) C-256-C-257.

12. High Toughened PSZ (Partially Stabilized Zirconia)

Takaki MASAKI* and Keisuke KOBAYASHI*

Abstract

PSZ (partially stabilized zirconia) is an excellent high-strength material toughened by martensitic transformation of metastable tetragonal particles.

This paper reviews recent research and development trend related to mechanical and thermal properties, micro structure and manufacturing method with various applications of PSZ.

Keywords : zirconia, partially stabilized zirconia, toughened ceramics, tetragonal zirconia polycrystal, microstructure, powder processing, ceramic scissors, stress induced transformation.

12.1 Introduction

Since the discovery of the PSZ transformation toughening mechanism achieved by Garvie et al.,[1] scientific research of ceramic toughening mechanism has continued with the aim to improve high strength and toughness properties of ceramics for use as mechanical structual materials.[2-5] The high strength and toughness of PSZ is considered to be due to the martensitic transformation from tetragonal (t) to monoclinic (m) phase which occurs near propagating cracks.[6] The toughening mechanism was introduced in a separate chapter by Dr. Faber, and is, therefore, not discussed here. The known representative PSZ is classified by composition into CaO-ZrO_2,[7] MgO-ZrO_2[8,9] and Y_2O_3-ZrO_2 systems, and by the formation method of tetragonal particles, into the precipitate type or the sintered polycrystalline type.[10,11]

The precipitate type PSZ is represented by MgO-ZrO_2 and CaO-ZrO_2 systems, and research results for modifying their properties by various heat treatments have been publicized. The high fracture toughness (K_{1c}) of Mg-PSZ has been given particular attention.[12]

On the other hand, the sintered polycrystalline type PSZ is represented by Y_2O_3-ZrO_2 system. It is also called the tetragonal zirconia polycrystal (TZP). Gupta et al.[13,14] first found that the tetragonal sintered body metastable at room temperature is obtained by microcrystallizing the sintered particles. Detailed studies on the stability of tetragonal particle in terms of Y_2O_3 composition, particle size of sintered body and density have since

---

* Technical Development Department, Toray Industries, Inc., 3-2-2, Sonoyama, Ohtsu, Shiga 520, Japan.

continued mainly by American and Japanese researchers.[15-17]

Y-TZP is characterized by nearly a 100 % tetragonal structure and very high bend strength.[18,19] The high strength and high toughness of Y-TZP is also considered to be caused by martensitic transformation as for Ca or Mg-PSZ.[15,20,21]

The present authors discuss recent research trends related to precipitate type and sintered polycrystalline type PSZs focusing on new development in mechanical properties and microstructure.

12. 2 Recent Progress in Mg-PSZ

12.2.1 Mechanical Properties

Recent attention has been given to the Mg-PSZ inelastic fracture phenomenon. Brittle ceramic strength is represented by the following equation, which complies with classic linear elastic fracture mechanics:

$$\sigma_f = \frac{1}{Y} \cdot \frac{K_{1c}}{C_f} \tag{12.1}$$

where, K_{1c}: fracture toughness (stress intensity factor)

C_f: critical crack size

Y: geometric constant (~ 1.1).

In this connection, Lankford,[22] Larsen[23] and Swain[24] made detailed experiments on the fracture behavior of Mg-PSZ. A Mg-PSZ stress-strain curve plotted by Larsen using the bending test method, from which the presence of inelastic behavior is confirmed, is shown in Fig. 12. 1. Later, Swain examined the relation between inelastic behavior and critical stress.

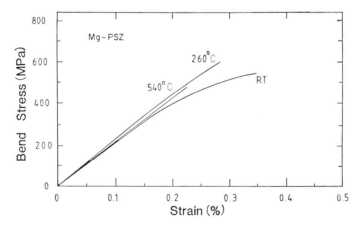

Fig. 12.1 Stress-strain curves in flexure for Mg-PSZ at three different temperature. [after D. C. Larsen[23]]

The critical stress in toughened zirconia is determined to be stress developed when the transformation from tetragonal to monoclinic structure is initiated. A stress field (σ_r) around a crack tip is determined from the following equation by applying the "Small Scale

Yielding" approximation:[25]

$$\sigma_r = \frac{K_{1c}}{2r} f(\theta) \tag{12. 2}$$

where, σ_r: stress field around crack tip

r: distance from crack tip

$f(\theta)$: a term depending on the direction from crack tip, when vertical, $f(\theta) = 1$.

In Eq.(12. 2), σ_r plays a function of r and K_{1c}. In case of $f(\theta) = 1$, when the transformation zone size (d) is given, the critical stress (σ_r^c) is determined.

Experimental methods such as TEM (transmission electron microscope), X-ray diffraction and Raman spectrometry for obtaining the d value have been proposed.[27-29] Swain determined $\sigma_r^c \sim 570$ MPa from Fig. 12. 2 by measuring $d \sim 70 \mu$m.

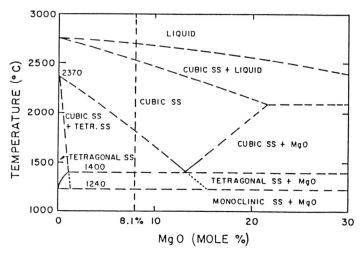

Fig. 12. 2 Phase diagram of the ZrO_2-rich portion of the $MgO-ZrO_2$ binary system. The dotted lines show the metastable extention of the solvuses defining the (c) + (t) region. [after C. F. Grain[33]]

On the other hand, the transformation zone size is given by the following equation, theoretically introduced by Evans et al.[6,30] with relation to K_{1c}:

$$K_{1c} = K_{1c}^m + \eta V_f \Delta V E d / (1 - \nu) \tag{12. 3}$$

where, K_{1c}^m: the inherent toughness of matrix containing transformed precipitates for particles

V_f: the volume fraction of transformable tetragonal phase (about 35% for 9.4 mol% Mg-PSZ)

ΔV: the volume dilation due to the transformation

E: Young's modulus (~ 200 GPa)

d: the transformation zone size

η: constant

ν: Poisson's ratio.

The value of η is calculated to be 0.2 to 0.50, based on the transformation stress field by Evans et al.[31]

Recently, Swain found that no correlation exists between the maximal values of strength (σ_f) and fracture toughness (K_{1c}) in Mg–PSZ.[32] The maximum value of σ_f is obtained at a PSM (polished surface monoclinic content) of 10%, and that of K_{1c} at a PSM of 20%. Swain proposed that a region exists in which strength is limited by critical stress when toughness is high (~ 12 MPa$\sqrt{\text{m}}$). In consideration of low toughness, he also noted a region where strength is limited by defects in compliance with Eq. (12. 1).

12.2.2 Microstructure

Characteristics of Mg–PSZ indicate that machanical and thermal properties are improved by controlling the microstructure through heat treatment. A group headed by CSIRO in Australia have been very active in this field.[9-11] As shown by the equilibrium phase diagram in Fig. 12. 3, a material containing 8 to 9.5 mol% of MgO is formed into a sintered body having a cubic grain size as coarse as 50 to 80 μm by sintering the material at temperatures as high as 1,700° to 1,800°C. Subsequently, during cooling after sintering, a tetragonal ZrO$_2$ precipitate is formed in a cubic matrix.[34] As for the precipitate, two methods are proposed: Ageing at temperatures as high as 1,300° to 1,500°C or temperatures lower than the eutectoid transformation point.[10,35,36] The recently proposed latter method, subeutectoid ageing, gives less coarsened particle size.

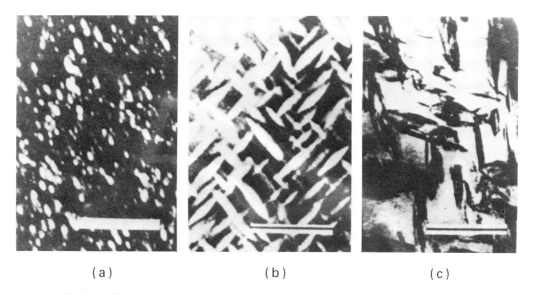

Fig. 12.3 Transmission electron microscope images of Mg–PSZ, (a) as-fired, dark field image, (b) aged at 1,420°C for 2 hours, and (c) for 4 hours. The latter figure is a bright field image. Bar length, 0.4μm. [after M. Marmach and M. V. Swain[10]]

The TEM observation of results using the former method is shown in Fig. 12. 4. Observations indicate that when exeeding critical size, the tetragonal precipitate loses coherency with the matrix and then changes from t to m phase during cooling. In

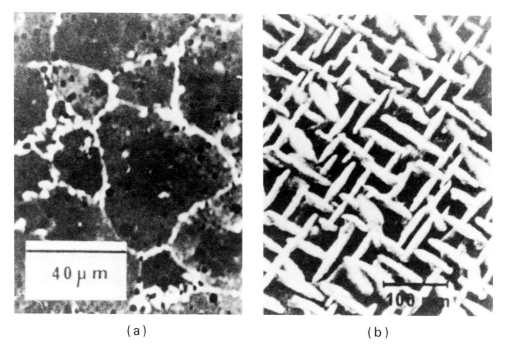

Fig. 12.4 Microstructure of Mg-PSZ after ageing at 1,100°C, (a) optical microstructure. The figure shows the development of a grain boundary monoclinic phase. (b) scanning electron microscope image of tetragonal precipitates in a cubic stabilized matrix. [after M. Marmach and M. V. Swain[10]]

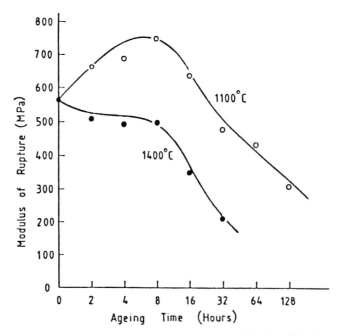

Fig. 12.5 Modulus of rupture variation of Mg-PSZ (9 mol% MgO) with heat treatment time at 1,100 and 1,400°C. [after M. Marmach and M. V. Swain[10]]

"sub-eutectoid" ageing, almost no precipitate is coarsened. Instead, a regulated defect fluorite structure is produced from considerable strain between the precipitate and matrix, initiating decomposition of the matrix phase along grain boundaries.[37] However, the heat treatment of "subeutectoid" ageing has not been sufficiently clarified due to its complexity. A change in microstructure before and after "sub-eutectoid" ageing is shown in Fig. 12.5. The lighter phase surrounding the particle in the figure is the monoclinic structure formed during decomposition.

Effects of ageing time in the two heat treatments are shown in Fig. 12.6. The figure shows that strength is greatly affected by heat treatment. Strength is considered to be directly caused by K_{1c} because the flaw size does not change by heat treatment.

In the future, number of findings on modifications of tetragonal precipitates in the precipitate type PSZ will be introduced, while diversified steel heat treatments continue to be developed.

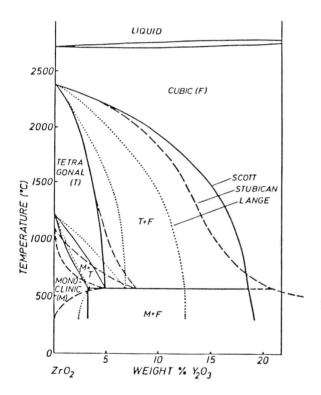

Fig. 12.6 Phase diagram in the ZrO_2 rich part of Y_2O_3 (Scott[39], Stubican et al.[40], Lange[41]). [after M. Rühle et al.[38]]

12.3 Recent Development in Y-TZP

12.3.1 Mechanical Properties

Recent 3 equilibrium phase diagrams of the ZrO_2-Y_2O_3 system in a composition range of small content of Y_2O_3 are shown in Fig. 12.7.[38,41] Each diagram shows the same form qualitatively, but different forms quantitatively. In any case, unknown matter is left in

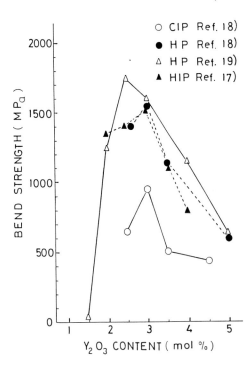

Fig. 12.7 Bending strength of Y-PSZ vs. Y_2O_3 content for three different processing methods. [after K. Kobayashi and T. Masaki et al.[43]]

the composition range of low Y_2O_3 content. Attempts to pursuit of high strength and high toughness of Y-TZP containing 2-3 mol% of Y_2O_3, as shown in the diagram, are being continued mainly in Japan and the production process are being improved.[18,42,43] The base is charecterized by very high bend strength (σ_f), e. g., 1.5 to 2.0 GPa.[17,19] The Y-TZP development process is shown in Fig. 12. 8. The σ_f values of sintered bodies produced by different molding and sintering methods, CIP (cold isostatic press), HP (hot press) and HIP (hot isostatic press), are given as a function of Y_2O_3 composition. For Y-TZP produced by the CIP method, the σ_f value lowers at less than 2.5 mol%, while for Y-TZP produced by the HP method, the σ_f value does not lower until it reaches a maximum value between 2.5 and 3.0 mol%. A Y-TZP having a σ_f value of not less than 1.75 GPa is obtained using high purity material.[19] Recent development in the production process using the HIP method instead of HP method, has made it possible to produce a high strength sintered body by making the particle size increasingly finer and denser.

The relation between the value of K_{1c} of a sintered body produced by the HP or HIP method and the composition of Y_2O_3 is shown in Fig. 12. 8. The value of K_{1c} is increased from 6 to 18 MPa$\sqrt{m}$ by the HIP method.[15,19,44]

Swain[24] measured the transformation zone size (d) using the X-ray diffraction method on a fractured surface of the HIP sintered body.

Results confirmed that K_{1c} is proportional to $\sqrt{d}$ as shown in Eq. (12. 1). While the Y-TZP particle size is not larger than 0.5 μm, the zone size exceeds particle size.

Swain examined Y-TZP fracture behavior in the same manner as for Mg-PSZ. He proposed that Y-TZP is limited by critical stress in compliance with Eq. (12. 2) rather than by the elastic fracture of Eq. (12. 1), as in Mg-PSZ.

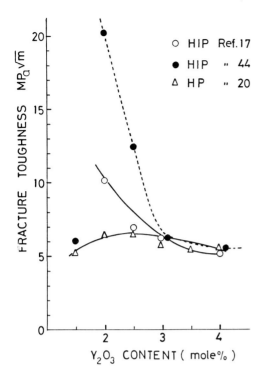

Fig. 12.8 Fracture toughness of hot-pressed and isostatically hot-pressed Y-PSZ vs. Y_2O_3 content. [after T. Masaki[19]]

Recently, a very high strength sintered body having a σ_f value of 2.3 GPa was obtained by the addition of twenty mol% of Al_2O_3 to ZrO_2.[45] Since the fracture toughness of the sintered body is low in this case, the bend strength is assumed to be greatly enhanced as a result of increased Young's modulus and decreased crack size (c_f) in Eqs. (12.1) and (12.3), caused by the addition of Al_2O_3.

12.3.2 Stability of Tetragonal Phase

Since this system consists of neanly a 100% tetragonal structure, the stability of a tetragonal phase is by far more important with respect to mechanical properties than for PSZ.

(a) Low Temperature Degradation of Y-TZP

It is reported that Y-TZP greatly degrades in strength when aged at low temperature (200 to 300°C), in contrast to Mg-PSZ which degrades by ageing over a wide temperature range.[16,46-49]

The strength degradation which is said to start at the surface, is thought to be due to the transformation from t to m structure. Changes in monoclinic fraction and thermal expansion property confirm this assumption. A microphotograph of Y-PSZ showing a change in microstructure by ageing is presented in Fig. 12.9. Change in microstructure caused by the transformation is clearly observed. In order to prevent strength degradation in the Y-TZP transformation, a number of studies are being made, including a measure to increase Y_2O_3 more than 3 mol% and to obtain fine particle size in ZrO_2.[42,46]

It was recently confirmed (see Fig. 12.10) that a dense Y-TZP with 0.5 μm particle size and greater than 6.07 g/cm³ density, prepared by the HP method using high purity

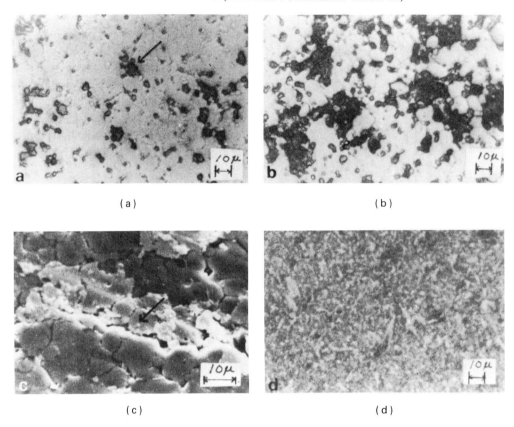

Fig. 12.9 Microstructural change of 4.5 mol% Y-PSZ. (a) optical micrograph of polished section, after sintering at 1,700°C. Monoclinic (M)-ZrO₂ is observed (arrow). (b) optical micrograph of (a) after at 300°C for 3,000 h. Note the segregation of M-ZrO₂. (c) SEM image of (a) after ageing at 300°C for 1,000 h. Microcrack is observed. (d) optical micrograph after sintering at 1,400°C. Note very fine grain size. [after K. Kobayashi et al.[16]]

powders, undergoes only minor changes in strength, monoclinic proportion and density even after long time ageing at temperatures between 200 and 500°C.[19]

(b) Degradation in Aqueous Solutions

Similar to the study of stability against ageing at low temperature, studies on degradation in aqueous solutions such as sulfuric acid, nitric acid, caustic soda and distilled water are being made.[50,51] A TEM photograph of 2.5 mol% Y-TZP taken before and after dipping in a 30% H_2SO_4 solution is shown in Fig. 12.11. The formation of twins and cracks along grain boundaries is found in many as a result of the dipping. The degradation is caused mainly by the martensitic transformation from t to m structure as in low temperature ageing. However, degradation is greatly accelerated in aqueous solution, in comparison with ageing.

Claussen[46] proposed a method in which Y-TZP is coated with a 10 to 40 μm thick Y_2O_3-rich ZrO_2 thin film, and Y-TZP is sintered in Y_2O_3 slurry in an attempt to prevent degradation.

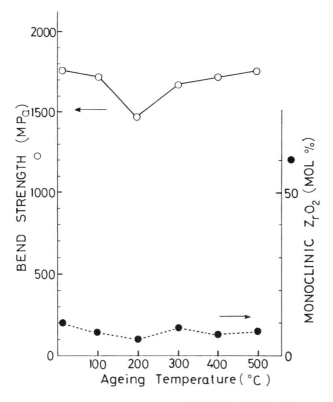

Fig. 12.10 Bend strengh or monoclinic ZrO_2 of 2.5 mol% Y_2O_3-PSZ hot-pressed at 1,500°C aftre ageing at various temperatures for 2,000 h. [after T. Masaki[19)]]

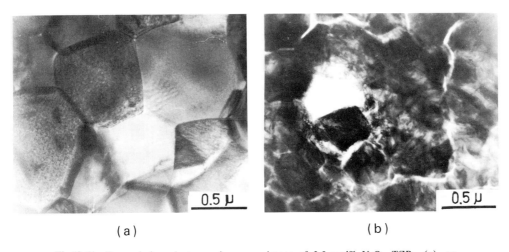

Fig. 12.11 Transmission electron microscope image of 2.5 mol% Y_2O_3-TZP, (a) as-received, (b) after dipped in 30% H_2SO_4 at 107°C for 96 h. [after K. Nakajima et al.[50)]]

12.3.3 Microstructure

Y-TZP is a single-phase structure consisting mainly of tetragonal structure, different from the fine tetragonal particles precipitated in a Mg-PSZ cubic matrix. The particle size is as fine as less than 1 μm. As shown in Fig. 12. 12, a HP sintered body has a dense structure such that almost no porocity is found both intergranularly and transgranularly.

Fig. 12.12 Transmission electron microscope image of 2.5 mol% Y_2O_3-TZP hot-pressed at 1,500°C. [after T. Masaki[19]]

A critical particles size is considered necessary in order to keep the metastable tetragonal structure stable at room temperature. However, the critical particle size has been the focus of active discussions by Chen et al. in recent years.[52,53]

A phase with complicated microstructure different from the above-stated single phase was recently found.[54-58] The microstructure is called rhombohedral,[58] modulated structure,[54] or tweed pattern[37] of the lamellar structure. As a result of detailed examinations in microstructure of the ZrO_2-Y_2O_3 alloys prepared by the Arc-melting method, Sakuma et al.[54] confirmed tha alloys containing two to three Y_2O_3 mol% having a structure with herring-bone type of metastable rhombohedral. An example of this herring-bone appearance is shown in Fig. 12. 13. The r-phase, initially found on a ground surface,[58] is considered to be a metastable phase since it assumes a structure strained from a cubic matrix. This complicated microstructure is thought to affect mechanical properties, but its nature and relation to metastable tetragonal phase have not yet been clarified.[54,59]

Fig. 12.13 Transmission electron microscope image of 3.0 mol% Y_2O_3-TZP. The figure shows Herring-bone structure. [after T. Sakuma[54]]

12.3.4 Powder Processing

More importance has been attached to powders for TZP than for PSZ. Synthesis or research and development of such powders, focusing on the ZrO_2-Y_2O_3 system are being carried out mostly by companies and research institutes in Japan.[60,65]

Characteristic values of Y-PSZ powders as in Fig.12.14, obtained by synthetic methods called wet processes, such as coprecipitation,[62] hydrolysis[63] or alkoxide methods are shown in Table 12.1. Besides these methods, wet processing includes pyrolysis and hydrochemical sythesis. The table shows that all these powders have high purity, fine particle size and large specific surface area. Besides these properties, the powder is required to be low in agglomerate. A dense Y-TZP of submicron particle size can be produced by sintering the powder having such characteristics at temperatures as low as 1,300 to 1,500°C.

Japan has pioneered industrial production of such high grade powders.

Table 12.1 Characteristics of PSZ powder.

Method	Chemical Analysis (wt %)						Crystal* Phase	Particle Size(by X-ray) (nm)	Surface Area (m²/g)
	Y_2O_3	Cl	SiO_2	Al_2O_3	Fe_2O_3	TiO_2			
Coprecipitation	7.1	0.05	0.08	0.03	0.003	0.04	T	20	20
Hydrolysis of Salts	5.3	0.2	0.02	0.002	0.002	0.003	0.8T +0.2M	20	15-25
Hydrolysis of Alkoxides	5.4	0.01	0.01	0.003	0.005		T		

* T : Tetragonal, M : Monoclinic

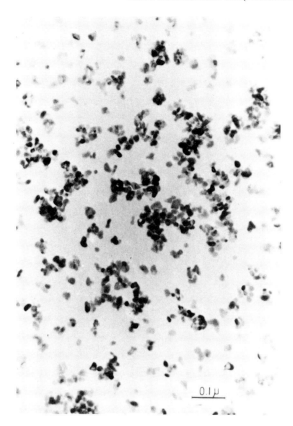

Fig. 12.14 Transmission electron microscope image of coprecipitated, fine zirconia powder.

12.3.5 Other Research

Besides Y-TZP, Y-PSZ single crystal is being studied mainly by a NAVY group.[71,73] The Y-PSZ single crystal produced by Skull-melting method is characterized by fine tetragonal precipitates and very high strength (~1,400 MPa).

It is thought that since the transformation toughening mechanism becomes ineffective at temperatures above 500°C, Y-TZP strength largely reduced. However, Y-PSZ single crystal can better retain its strength because of crack-precipitate interaction and the dispersion strengthening mechanism.[71]

12.4 Application

Companies primarily in Japan and Australia are vigorously developing applications to best utilize the excellent properties of Y-TEP, such as high strength and high toughness.[74-82]

Areas under the study for Y-TZP applications include (1) cutlery such as scissors (Fig. 12.15),[77] (2) grinding media, (3) sliding parts such as roller guides and (4) bioceramics and electronic ceramic components and diesel engine parts.

Mg-PSZ is used in such fields as (1) extrusion die[82] (Fig. 12.16) and (2) diesel engine parts.[74-76,78,81]

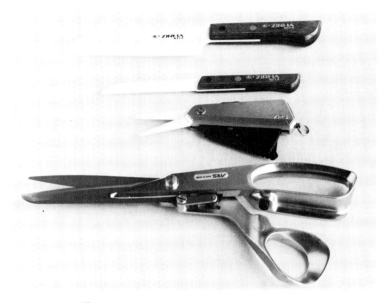

Fig. 12.15 Tape and textile scissors, knives, cutters.

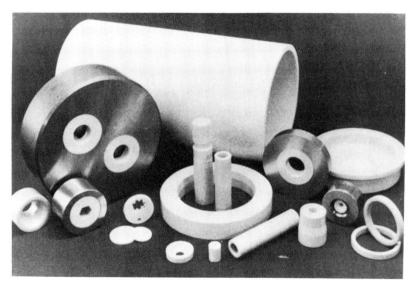

Fig. 12.16 An assortment of "Nilsen's PSZ" partially stabilized transformation toughened zirconia components.

PSZ and TZP are greatly improved in brittleness, which was the greatest disadvantage in ceramics, and have excellent characteristics such as low friction coefficient and surface smoothness which are not found in silicon nitride, silicon carbide or alumina.

Thus PSZ and TZP are promising for use in mechanical functional components or mechanical structural materials, comparable or superior to other types of ceramics mentioned above.

12.5 Conclusion

The study of PSZ has continued for 10 years since the discovery of the transformation toughening ZrO_2 by Gravie in 1975. Since then, Mg-PSZ has been actively researched and developed in Australia, America and Germany, while Y-TZP has been researched mainly in Japan. Particularly in Japan, attempts have focused on the production process to achieve high Y-TZP strength and toughness.

The main feature of Mg-PSZ is high toughness (K_{1c}: ~14 MPa$\sqrt{m}$), while Y-TZP features high strength (σ_f: ~1,750 MPa) and high toughness (~18MPa$\sqrt{m}$): both have excellent mechanical properties not obtained from other ceramics. The toughening mechanism generating high strength is being actively studied in America. The toughening is attributed to the so-called stress-induced transformation in metastable tetragonal particles present in Y-TZP or Mg-PSZ which transform in a martensitic manner under external stress field. On the other hand, studies of the microstructure with TEM are being promoted in order to observe the toughening mechanism. Recently, the presence of a complicated microstructure called the modulated structure was confirmed. The structure is drawing attention as the fourth structure besides the cubic, tetragonal and monoclinic structure in ZrO_2. The relation between microstructure and high toughness is being investigated, and the study of the relation will clarify the presence of a toughening mechanism other than t to m transformation.

In addition to the above-stated fundamental research, future PSZ development is expected to produce an excellent ZrO_2 toughened by the addition of third component (CeO_2, Al_2O_3, etc.) other than MgO or Y_2O_3. In parallel, production process will be studied, such as powder synthesis, molding and sintering, especially the production of a sintered body having dense and fine particle size mainly by the HIP method. In order for PSZ to be used as structural components for engines, and for promising applications other than cutlery and die in the future, it is necessary to collect more data on material and practical properties, as well as to improve PSZ heat resistance at high temperature.

References

1) R. C. Garvie, R. H. J. Hannink and R. T. Pascoe : Nature (London), 258 (1975) 703.
2) N. Claussen : Advances in Ceramics, Vol. 12 (1984) 325-351.
3) D. L. Porter and A. H. Heuer : J. Am. Ceram. Soc., **60**, 3-4 (1977) 183-184.
4) N. Claussen : J. Am. Ceram. Soc., **61**, 1-2 (1978) 85-86.
5) T. K. Gupta : Fracture Mechanics of Ceramics, Vol. 4, ed. by R. C. Bradt, D. P. H. Hasselman and F. F. Lange, Plenum Publishing Corp. (1978) 877-889.
6) R. McMeeking and A. G. Evans : J. Am. Ceram. Soc., **65**, 5 (1982) 242-245.
7) R. C. Garvie, R. H. J. Hannink, R. R. Hughan and N. A. Mckinnon : J. Aust. Ceram. Soc., **13**, 1 (1977) 8-11.
8) R. H. J. Hannink, K. A. Johnston, R. T. Pascoe and R. C. Garvie : Advances in Ceramics, Vol. 3 (1981) 116.
9) R. H. J. Hannink and M. V. Swain: J. Aust. Ceram. Soc., **18** (1982) 53.
10) M. Marmach and M. V. Swain : Proc. Int. Conf. on Ceramics for Engines, Jpn. Ceram. Soc. (1984) 651.
11) M. V. Swain, R. H. J. Hannink and R. C. Garvie : Fracture Mechanics of Ceramics, Vol. 6, ed. by R. C. Bradt, A. G. Evans, D. P. H. Hasselman and F. F. Lange, Plenum Publishing Corp., New York (1982) 339-353.
12) M. V. Swain and R. H. J. Hannink: Advances in Ceramics, Vol. 12 (1984) 225-239.
13) T. K. Gupta, F. F. Lange and J. H. Bechtold: J. Mater. Sci., **13** (1978) 1464-1470.

14) T. K. Gupta, J. H. Bechtold, R. C. Kuznicki, L. H. Cadoff and B. R. Rossing : J. Mater. Sci., **12** (1977) 2421-2426.
15) F. F. Lange : J. Mater. Sci., **17** (1982) 225-234.
16) K. Kobayashi, H. Kuwajima and T. Masaki : Solid State Ionics, **3/4** (1981) 489-493.
17) K. Tsukuma, Y. Kubota and T. Tsukidate : Advances in Ceramics, Vol. 12 (1984) 382-390.
18) T. Masaki and K. Kobayashi : Jpn. Ceram. Soc. Meeting (1981) 2, 3.
19) T. Masaki : Jpn. Ceram. Soc. Meeting (1984) 455.
20) F. F. Lange : J. Mater. Sci., **17** (1982) 235-246.
21) F. F. Lange : J. Mater. Sci., **17** (1982) 255-263.
22) J. Lankford : J. Am. Ceram. Soc., **66** (1983) C-212.
23) D. C. Larsen : IIT Research Institute, Tech. Report, No. 11 (1981).
24) M. V. Swain : Acta Metall., **33**, 11 (1985) 2083-2091.
25) B. R. Lawn and T. R. Wilshaw : Fracture of Brittle Solid, Ch. 3., Cambridge University Press (1975).
26) M. V. Swain : J. Am. Ceram. Soc., **68**, 4 (1985) 97-99.
27) D. R. Clarke and F. Adar : J. Am. Ceram. Soc., **65**, 6 (1982) 284-288.
28) R. C. Garvie, R. H. J. Hannink and M. V. Swain : J. Mater. Sci. Lett., **1** (1982) 437-440.
29) T. Kosmac, R. Wagner and N. Claussen : J. Am. Ceram. Soc., **64**, 4 (1981) C72-73.
30) B. Bubiansky, J. W. Hutchinson and J. C. Lambropoulos : Int. J. Solid Struct., **19** (1983) 337-355.
31) J. C. Lambropoulos : Ph. D. Thesis, Harvard University (1984).
32) M. Marmach and M. V. Swain : J. Aust. Ceram. Soc. (1984) (in press).
33) C. F. Grain : J. Am. Ceram. Soc., **50**, 6 (1967) 288-90.
34) D. L. Porter and A. H. Heuer : J. Am. Ceram. Soc., **62** (1979) 298-305.
35) R. H. J. Hannink : J. Mater. Sci., **18** (1983) 457-470.
36) R. H. J. Hannink and R. C. Garvie : J. Mater. Sci., **17** (1982) 2637-2643.
37) R. H. J. Hannink : J. Mater. Sci., **13** (1978) 2487.
38) M. Rühle, N. Claussen and A. H. Heuer : Advances in Ceramics, Vol. 12 (1984) 352-370.
39) H. G. Scott : J. Mater. Sci., **10**, 9 (1975) 1527-1535.
40) V. S. Stubican, R. C. Hink and S. P. Ray : J. Am. Ceram. Soc., **61**, 1-2 (1978) 17-21.
41) F. F. Lange : Fracture Mechanics of Ceramics, Vol. 6, ed. by R. C. Bradt, A. G. Evans, D. P. H. Hasselmann and F. F. Lange, Plenum Publishing Corp., New York (1983) 255-274.
42) T. Tsukuma, Y. Kubota and K. Nobugai : Yogyo Kyokai-Shi, **92**, 5 (1984) 233-241.
43) K. Kobayashi and T. Masaki : Bull. Jpn. Ceram. Soc., **17** (1982) 427-433.
44) T. Masaki : Jpn. Ceram. Soc. Meeting (1985) 665.
45) K. Tsukuma, K. Ueda and T. Tsukidate : Jpn. Ceram. Soc. Meeting (1984) 119.
46) N. Claussen, H. Schubert and M. Rühle : to be presented at the 84th Annual Meeting of the Am. Ceram. Soc., Pittsburgh (1984).
47) T. Sato, O. Ohtani and M. Shimada : J. Mater. Sci., **20** (1985) 1466-1470.
48) W. Watanabe, S. Iio and I. Fukuura : Advances in Ceramics, Vol. 12, Am. Ceram. Soc. (1984).
49) M. Matsui, T. Soma and I. Oda : Advances in Ceramics, Vol. 12, Am. Ceram. Soc. (1984).
50) K. Nakajima, K. Kobayashi and M. Murata : Advances in Ceramics, Vol. 12, Am. Ceram. Soc. (1984).
51) T. Sato and M. Shimada : Jpn. Ceram. Soc. Meeting (1984) 461.
52) I.-W. Chen and Y.-H. Chiano : Advances in Ceramics, Vol. 12, Am. Ceram. Soc. (1984) 33-45.
53) I.-W. Chen and Y.-H. Chiano : Acta Metall., **31** (1983) 1627-1638.
54) T. Sakuma : Jpn. Ceram. Soc. Meeting (1984) 123 ; J. Mater. Sci., to be published.
55) T. Watanabe, K. Urabe, H. Ikawa and S. Udagawa : Jpn. Ceram. Soc. Meeting (1984) 463-466.
56) Y. Moriyoshi, T. Ikegami, S. Matsuda, H. Yamamura and A. Watanabe : Jpn. Ceram. Soc. Meeting (1984) 131.
57) N. Ishizawa, A. Saiki, H. Oka, N. Mizutani and M. Katou : Jpn. Ceram. Soc. Meeting (1984) 129.
58) H. Hasegawa : J. Mater. Sci. Lett., **2**, 3 (1983)

59) P. G. Valentine, R. D. Maier and T. E. Mitchell : National Aeronuatics and Space Administration Grant No. NSG-3252 (1981).
60) J. Sakai : J. Jpn. Ceram. Soc., **17** (1982) 454-458.
61) T. Tsukuma and K. Tsukidate : J. Jpn. Ceram. Soc., **17**, 10 (1982) 816-822.
62) K. Haberko : Ceramurgia Int., **5**, 4 (1972) 148.
63) K. S. Mazdiyasni et al. : J. Am. Ceram. Soc., **50** (1967) 532.
64) Zirconia Ceramics 1, ed. by S. Somiya, Uchida Rokakuho Pub., Co., Ltd. (1983) 1-19, 21-28, 31-34, 45-60.
65) Ceramic Science and Technology at the Present and in the Future, ed. by S. Somiya, Uchida Rokakuho Pub., Co., Ltd. (1982) 501-514.
66) C. E. Scott et al. : Am. Ceram. Soc. Bull., **58** (1979) 587.
67) W. H. Rhodes : AFML-TR-70-209 (Sept. 1970).
68) W. H. Rhodes : J. Am. Ceram. Soc., **64**, 1 (1981) 14-19.
69) P. H. Rieth, J. S. Reed and A. W. Heumann : Am. Ceram. Soc. Bull., **55**, 8 (1976) 717-721, 727.
70) W. H. Rhodes : J. Am. Ceram. Soc., **64**, 1 (1981) 20-22.
71) R. P. Ingel : Ph. D. Thesis, Catholic University of America (1982).
72) R. P. Ingel, D. Lewis, B. A. Bender and R. W. Rice : J. Am. Ceram. Soc., **65**, 9 (1982) C150-152.
73) R. P. Ingel, D. Lewis, B. A. Bender and R. W. Rice : Advances in Ceramics, Vol. 11 (1984) 408-414.
74) R. Kamo and W. Bryzik : "Adiabatic turbocompound engine performance prediction", SAE Paper 780068 (1978).
75) M. E. Woods and I. Oda : "PSZ ceramics for adiabatic engine components," SAE Technical Paper Series 820429 (1982).
76) M. Marmach and D. Servent : SAE Paper 830318 (1983) 64-72.
77) T. Matsumoto : Japan F. C. Report, **2**, 12 (1984) 24-28.
78) H. Irokawa, I. Oda and N. Yamamoto : Japan F. C. Report, **2**, 12 (1984) 1-6.
79) R. C. Garvie : Advances in Ceramics, Vol. 12 (1984) 465-479.
80) R. C. Garvie, C. Urbani, D. R. Kennedy and J. C. McNeuer : J. Mater. Sci., **19** (1984) 3224-3228.
81) S. Robb : "Cummins successfully test adiabatic engine," Bull. Am. Ceram. Soc., **62**, 7 (1983) 755-756.
82) S. T. Gulati, J. D. Helfinstine and A. D. Davis : Bull. Am. Ceram. Soc., **59**, 2 (1980) 211-219.

13. Ceramic Application for Automotive Components

Shigetaka WADA*

Abstract

This paper discusses features of various automotive ceramic components for electronic and structural uses. The adiabatic diesel engine and the gas turbine with high turbine inlet temperature are regarded as answers to demands for highly efficient internal combustion engines. Ceramic components for such engine systems have been developed in major countries of the world.

Structural ceramic components are discussed from the standpoint of feasibility, reliability and cost. It is suggested that high cost is the largest obstacle and the key process to reduce manufacturing cost of ceramic components is post-sintering machining. The introduction of structural ceramic components to engines is considered a step toward the all-ceramic engine which will be a major contribution to the automobile industry of the future.

Keywords : ceramic, automotive components, feasibility/reliability/cost.

13.1 Introduction

The automotive industry in Japan established its manufacturing expertise and mass production facilities in the 1950's and 1960's. The development of engine technology on exhaust emission control started in the early 1970's. The oil shortage in 1973 necessitated improvements in fuel economy in the automotive industry throughout the world. Japanese cars acquired the reputation for quality at competitive cost at an international level in the late 1970's.

Few ceramic components were applied to cars of the 1950's except for window glasses and spark plugs. In the past two decades, many components utilizing various ceramic materials have been introduced into cars.

Favorable general reviews have been published on ceramic materials and their properties, ceramic components and their functions, and ceramic fabrication methods,[1-9] including an excellent review on the engineering aspects of available materials for engines and the use of ceramics in gas turbines, internal combustion engines and tribological applications.[10] There are also current reviews on the development of high-temperature

---

* Toyota Central Research & Development Laboratories, Inc., Nagakute, Aichi 480-11, Japan.

ceramic and composite materials for advanced heat engine components, particularly turbochargers, and gas turbines.[11-13]

Since there are several useful reviews on the ceramic applications for automotive components as mentioned above, similar topics are avoided in this article. This paper describes and discusses the following:

–Features of automotive components utilizing electronic ceramics and structural ceramics.
–Major obstacles to overcome to realize the ceramic engine.

13.2 Electronic Ceramics Versus Structural Ceramics in Automotive Components

Major components utilizing ceramics in Toyota's mass production cars are shown in Fig. 13.1.[14] Ceramics used in automotive components are divided into two groups, electronic ceramics and structural ceramics, and are classified according to the material and its properties as shown in Table 13.1. It can be seen that there are several different features between a component utilizing electronic ceramics and that which utilizes structural ceramics.

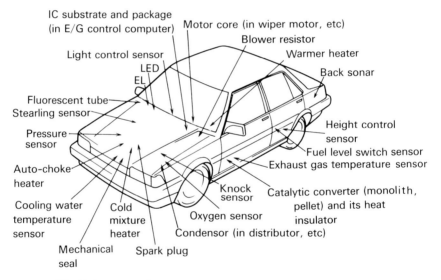

Fig. 13.1 Automotive components utilizing ceramics in Toyota car.

(1) Thirty-eight components are noted in Table 13.1. Electronic ceramic components are greater in number (twenty five), and more varied in properties and materials than structural ceramic components. It is easily understood that most electric and electronic ceramic properties known today have already been applied in the automotive components.

(2) Most electronic ceramic components have been used in many car models by automotive manufacturers. On the other hand, some structural ceramic components except for the components related to exhaust gas cleaning, e. g., glow plugs, pre-combustion chambers and rocker arm chips, are used for only one or two car models.

(3) Ceramics used in electronic components are light in weight and small in size.

Table 13.1 Automotive components utilizing ceramics.

Classification	Property	Material	Component
Electronic Ceramics	NTC thermistor	Fe_3O_4, Mn_2O_3, NiO $Al_2O_3 \cdot Cr_2O_3$, ZrO_2-Y_2O_3	Fuel level switch Cooling water temperature sensor Exhaust gas temperature sensor
	Positive temperature coefficient of resistance	Semiconductive $BaTiO_3$	Cold mixture heater (Riser type, Pipe type) Auto-choke heater, Blower resistor
	Ion conductivity	ZrO_2-Y_2O_3	Oxygen sensor Lean mixture sensor
	Piezoelectricity	Pb(Zr, Ti)O_3, etc.	Knock sensor, Electric buzzer Back sonor, Rain drop sensor Oil level sensor
	Ferroelectricity	$BaTiO_3$	Condensor (in distributor, etc.)
	Ferromagnetism	$MnO \cdot Fe_2O_3$, Fe_3O_4, $CoO \cdot Fe_2O_3$	Motor core (in Wiper motor, etc.)
	Semiconductivity	Si, $MgCr_2O_4$, TiO_2, SnO_2	Pressure sensor Dew sensor, Humidity sensor
	Varistor	SiC, ZnO_2	Surge absorber (in electronic circuit)
	Photoconductivity	CdS, ZnO, PbS	Light control sensor
	Electroluminescence	ZnS	EL display
	Electrofluorescence	ZnO	Fluorescent tube
	LED	GaP	LED display, Height control sensor Stearling sensor
	Electric insulation	Al_2O_3	IC-substrate and package
Structural Ceramics	Mechanical strength Electric insulation	Al_2O_3	Spark plug
	Transparency	Na_2O-CaO-SiO_2 glass	Window glass, Light valve
	Wear resistivity	Al_2O_3, PSZ Si_3N_4	Mechanical seal Rocker arm chip
	Low thermal expansion Chemical stability	Cordierite ($2MgO \cdot 2Al_2O_3 \cdot 5SiO_2$) Al_2O_3	Catalytic converter (Monolith type, Pellet type)
	Large specific surface area	γ-Al_2O_3	γ-coat on monolithic catalytic conveter
	High refractoriness Heat insulation	Al_2O_3-SiO_2 fiber	Heat insulator for catalytic converter, etc.
	High strength Thermal shock resistance	Si_3N_4	Glow plug pre-combustion chamber
	Wear resistivity	$Al_2O_3 \cdot SiO_2$ fiber	Piston ring (FRM)

Ceramics comprise less than 10% of the weight of most components. The absolute weight of ceramics is less than 1 gm. for many sensors and about 20 gm. for partially stabilized zirconia (PSZ) in the oxygen sensor; PSZ is considered to be a heavy ceramic part in electronic ceramics.

In contrast, structural ceramics used in the components are heavy in weight and large in volume on the average, even excluding the exceptional case of window glasses (20 kg total weight). Monolithic catalytic converters weigh about 0.5 kg with a volume of 1.0 - 1.5 l. A pre-combustion chamber of silicon nitride which has been applied to commercial cars weighs about 20–30 gm. per piece.

(4) Most electronic ceramic materials and components had been used in other industries before application in the automotive industry. For example, NTC thermistors have been used for measuring instruments such as temperature sensors, while piezoelectric ceramics have been applied in the fishing industry as sonors, in communications as high frequency filters and in home appliances as piezo-lighters. Semiconductive barium titanate ceramics with a positive temperature coefficient of resistivity (PTC thermistor) were invented in the early 1950's and have been used in home appliances as color TV degaussing elements, and as heating elements for rice cookers, clothes dryers, and hair dryers prior to application as an automatic choke heater and a cold mixture heater (a heater which enhances fuel evaporation at cold start). PSZ is an exceptional electronic ceramic material. Its mechanical and electrical properties were improved largely for adaptation to the oxygen sensor. The main effort in the development of the electronic ceramic components has been to design components durable enough to withstand severe conditions, such as high temperature, high humidity and vibration, utilizing already known properties of ceramics.

On the contrary, most structural ceramic materials and components were not used in other industries until they were used as automotive components or were improved to be fitted for automotive components.

The spark plug was first used in 1860 for the two cylinder engine invented by E. Lenoier, as an indispensable component in the ignition system of an internal combustion engine. Since then, various ceramic materials such as feldspar, mica, steatite, silimanite, mullite, spinel and cordierite have been used in spark plugs. An alumina ceramic "sinterkorund" was first developed in 1932 in Germany.[16] Since then, various ceramic materials for spark plugs have been substituted by alumina ceramic. Research on an alumina ceramic for spark plugs has formed solid background for an alumina substrate and a package in the electronics industry.

Monolithic catalytic converters having a honeycomb-like structure are composed of thin walls of approximately 0.1-0.3 mm thickness and 1-3 mm channel-width. The monolithic converter material should possess low thermal expansion, because it is installed at the immediate downstream of the exhaust manifold and is exposed to high temperatures and severe thermal cycling. Presently, the monolithic converter is made from a cordierite ceramic with a low thermal expansion coefficient of 0.7-1.0 $\times$ 10^{-6}/°C. This low thermal expansion coefficient depends on the degree of preferred orientation of the c-axis of the cordierite crystal parallel to the thin wall which results from the orientation of talc, one of the raw materials, during extrusion forming and topotactic crystal growth of cordierite crystal from the talc crystal. This mechanism and production technique to obtain low thermal expansion monoliths were initially invented for this application.[17]

(5) As sensors, electronic ceramics have mainly contributed to the quantitative improvement of engine function through optimal control by electronics. On the other hand, structural ceramics have added several qualitative functions to the engine system.

Electronic ceramics and structural ceramics have contributed to the progress in automotive performance utilizing characteristics mentioned above. Important technical aspects for improvement or development in the future are as follows:[17-19]

—Improvement in fuel economy

Table 13.2 Automotive components utilizing ceramics development stage.

Classification	Property	Material	Component	Application
Electronic Ceramics	Semiconductivity	TiO_2	Oxygen sensor	A/F control
	Ion conductivity	$\beta\text{-}Al_2O_3$	Na-S battery	Electric car
Structural Ceramics	High refractoriness	Cordierite ($2MgO \cdot 2Al_2O_3 \cdot 5SiO_2$)	Particulate filter	Diesel suit filter
	Low thermal expansion Low thermal conductivity	LAS $Al_2O_3 \cdot TiO_2$	Exhaust port liner Exhaust manifold	
	Light specific weight High strength	Si_3N_4, Sialon, SiC	Rotor	Turbocharger Turbo-compound E/G
	Thermal shock resistance	SiC fiber reinforced LAS, etc.	Housing	
	Wear resistance Low friction High strength Light specific weight	Si_3N_4, Sialon, SiC	Pushrod chip, valve guide Cum, Bushing Socket Piston ping Connecting rod Bearing for connecting rod, etc.	Reciprocating E/G (Minimum friction E/G)
	High strength Moderate thermal expansion Low thermal conductivity	PSZ, ZrO_2- coating Si_3N_4	Cylinder liner Cylinder head hot plate Piston head Exhaust valve and valve sheet, etc.	Adiabatic diesel E/G
	Low thermal expansion High refractoriness	LAS, AS	Regenerator Flow separator housing	
	High strength High refractoriness	SiC and SiSiC Sialon, Si_3N_4	Combustor, Scroll, Rotor Stator, Turbine shroud. etc.	Gas turbine
	High strength Low thermal conductivity	PSZ	Heat insulator	
	Low thermal conductivity	Al_2O_3-SiO_2 fiber		

LAS : lithium aluminium silicate.
AS : aluminium silicate.
SiSiC : silicon infiltrated SiC.
PSZ : partially stabilized zirconia.

-Improvement in engine fuel effieciency
-Reduction of inertia weight
-Improvements in aerodynamics
-Development of low loss transmission
-Elimination of other mechanical loss
-Development of high-powered engine
-Mitigation of noise
-Countermeasures for alternative fuel

Electronics and new materials are likely to provide the most effective means for improvements in these areas.

Automotive components utilizing ceramics, which are under development or expected to be in the market, are shown in Table 13. 2. It is noted that almost components listed in Table 13. 2 are structural ceramic components. It is seen from Tables 13. 1 and 13. 2 that most electronic ceramics with promising properties for automotive components have already been applied to cars in the market. This implies that new sensors or electronic ceramic components cannot be expected to be introduced from other industries. Automotive manufacturers or automotive component manufacturers need to conduct their own basic research within electronic ceramics.

13. 3 Three Obstacles to Practical Use of Structural Ceramics

Several ceramic engine components are in practical use, as shown in Table 13. 1, after more than ten years of research and development efforts in the field of structural ceramics. This does not mean an all-ceramic engine but the partial substitution of metallic parts with ceramic components. The introduction of ceramic components to engines, however, is a step toward the all-ceramic engine which will contribute to the future automotive industry.

Three obstacles which must be overcome to realize the all-ceramic engine are:

Feasibility   Higher engine efficiency is expected using an exhaust energy utilization system such as a turbocharger and turbocompound or by raising the turbine inlet temperature in the gas turbine engine. Initially, this higher engine efficiency must be demonstrated by engine tests and the ceramic components must prove durable in all driving conditions.

Reliability   The feedback process to assure reliability of ceramic components on the rig test and engine test, as shown in Fig. 13. 2, was proposed when the "Brittle Material Design" program, the first big project intended to develop ceramic components in a gas turbine engine, was planned.[20] Such development feedback processes are necessary due to the inherent brittleness in ceramics and the mechanical reliability of ceramic components cannot be evaluated definitely without applying practical stress. Although the study on fracture and life prediction in ceramics, non-destructive evaluation (NDE) technology, and basic research on ceramic materials have made good progress in the past ten years, nearly the same feedback process as shown in Fig. 13. 2 is required even at present for the application of ceramics in engine components. Furthermore, durability data on oxidation, corrosion and erosion for various materials and atmospheric conditions must be accumulated.

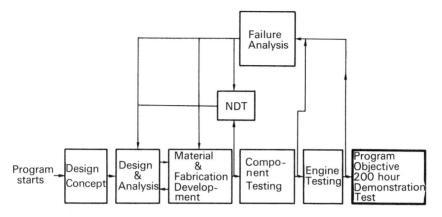

Fig. 13.2 ARPA Programme, "Brittle Materials Design, High Temperature Gas Turbine"; flow chart of major tasks and iterative development. [after A. F. McLean[20]]

Cost The price for automotive components is generally low, as will be described later. Although ceramic engines may prove to be highly efficient, they will be used practically only if they can be manufactured at reasonable cost.

"Cost" appears to be the greatest obstacle to overcome in comparison with problems of feasibility and reliability. Machining cost is one of the major factors which increase total production cost.

13.3.1 Feasibility and Reliability

There are several ceramic components which are used under severe thermal, mechanical and chemical ambience.

The PSZ element for the oxygen sensor is installed mechanically in a metal housing and can endure high temperatures of 900–1,000 °C, thermal shock of more than 50 °C/sec and vibration of 50 G. About 500,000 oxygen sensors, used under the severe conditions described above, are now produced monthly in Japan.

The monolithic catalytic converter made of cordierite ceramic withstands temperatures of more than 900–1,000 °C and thermal spalling tests of 600–700 °C temperature differences. Approximately 100 million pieces have been produced in the world.

The Si_3N_4 diesel glow plug, developed by Isuzu and Kyocera, can withstand rapid temperature rise up to 1,000 °C within three seconds.[21] It was reported that about 1.4 million pieces had been produced by the end of 1984.

Si_3N_4 pre-combustion chambers in indirect diesel engines have been employed in Toyota and Isuzu commercial cars. The tensile stress which acts in the ceramic chamber as the sum of thermal stress and mechanical stress is only about 10 kgf/mm² because of the optimization of shapes based on stress analysis by FEM. The three-point bending strength of the injection molded and conventionally sintered Si_3N_4 for pre-combustion chamber in conformity with the JIS-standard R-1601 is 75 kgf/mm²,[22] which can be converted to the tensile strength of about 40 kgf/mm². It is understood that the Si_3N_4 pre-combustion chamber is designed with a fair margin of strength. It is reported that the number of pre-combustion chambers used in Isuzu cars has totaled 0.14 million.

The exhaust valve is used under severer conditions than all other adiabatic diesel engine components. The maximum load of 1,360 kg, corresponding to a maximum stress of 30.6 kg/mm², is imposed on one exhaust valve. It was reported that the Si_3N_4 exhaust valve and valve sheets were tested successfully on an engine for 500 hrs at 6,000 rpm by Isuzu;[23] the exhaust valve withstood 1.5×10^6 stress cycles.

Utilizing the five kinds of components mentioned above as automotive components had been considerably difficult until several years ago. However, today PSZ elements for oxygen sensors, cordierite monolithic catalytic converters, Si_3N_4 glow plugs and pre-combustion chambers, can be used without serious technical trouble. These positive advancements suggest that many components for heat engine systems which are still difficult to apply in commercial cars as well as the all-ceramic engine in itself will prove to be technically feasible and reliable in the near future.

13.3.2 Cost

The price per unit weight of cars is considerably low in comparison with other familiar goods as shown in Fig. 13. 3.[24] An example of the raw materials which comprise passenger cars is shown in Fig. 13. 4.[25] The relation between the price per gram and total production in tons per year in Japan for major automotive materials and related materials is shown in Fig. 13. 5.

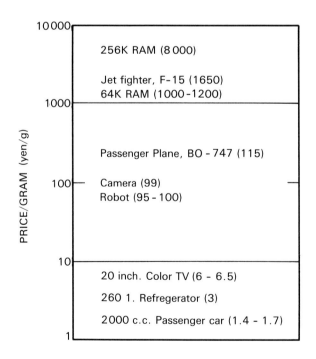

Fig. 13. 3 The price per unit weight for several goods.

Since the retail price of cars is 1.4–1.7 yen/gm., the production cost of car will be nearly 1.0 yen/gm. The major automotive materials are iron, steel, aluminium, sheet glass and plastics as shown in Fig. 13. 4, and their price is only 0.07–0.3 yen/gm. The price per

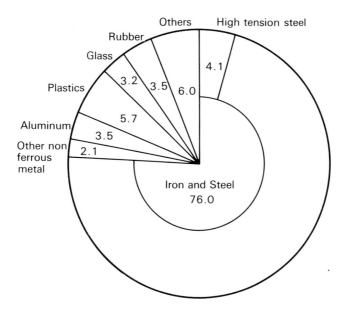

Fig. 13.4 The fraction of materials composing Japanese passenger car (wt. %, 1983).

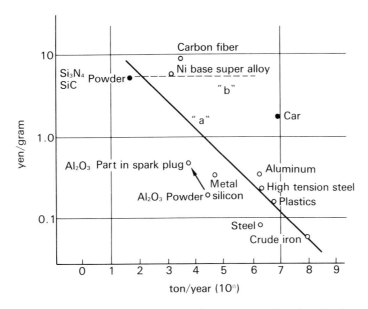

Fig. 13.5 The relation between the price per gram and total production in tons per year in Japan for major automotive materials.

gram of these materials and their production in weight per year is correlated as line "a" in Fig. 13. 5. A car is composed of 20,000–30,000 components and not all components and materials are assessed by price per weight. However, when the amount of some component or material used in one car exceeds one percent (approximately 10 kg), it must be accepted that the cost per weight of that component or material is one of the important assessment criteria for adoption.

Prof. K. H. Bowen proposed that the relation between the manufacturing cost of ceramic components and the potential amout of components used in the reciprocating engine could be predicted as shown in Table 13. 3. The important aspect of this table is that the unit amount of components having the potential for use is not the "total weight of production" but "mass/engine." That is, the total weight and/or the number produced is one of the essential factors affecting cost from the standpoint of component makers. On the contrary, the cost of the component in one car is important for automotive manufacturer, even when the component is applied to only one car model.

Table 13. 3 Ceramic components in reciprocating engine. [after H. K. Bowen[26)]]

Manufacturing Cost of Components	Mass/Engine
130–855 S/kg	0.1–0.4 kg
85–40	0.4–2.0
40–25	2.0–10.0

Si_3N_4 glow plug, turbo, bearing, turbo housing, cutting tools, much larger applications.
(taken from the presentation at twelfth automotive material conference "ceramics in engine".)

The total amount of ceramic components in weight per adiabatic turbocompound engine or gas turbine engine for passenger cars is estimated to be 10–20 kg. The 10–20 kg weight is equivalent to 1 % of a standard passenger car weight. The allowable upper limit cost of ceramic components for all ceramic engines with reference to Table 13. 3, therefore, seemed to be 6 yen/gm. which is shown by the dotted line "b" in Fig. 13. 5.

The price of raw powders for Si_3N_4, SiC and PSZ, the most prominent ceramics for the components of a heat engine, is approximately 5–10 yen/gm. and their total production does not amount to 100 ton/year for any powders in Japan today. When adiabatic engines and/or gas tubines prove to be feasible technically and the all-ceramic engine is installed for about 2 % of the total automotive production in Japan (200,000 cars/year), the total weight of ceramic components will amount to 2,000 ton/year. The practicality of producing ceramic components at less than 6 yen/gm. on that scale of production is questionable.

If it is anticipated that the price of raw powders decrease down in relation to line "a" in Fig. 13. 5, 2 yen/gm. price expected when production quantity amounts to 2,000 ton/year. Si_3N_4 powders for engineering ceramics are mainly synthesized from the nitridation method of silicon metal. In Japan almost 100 % of silicon metal has been imported and its price is 0.3 yen/gm. Japanese manufacturers of ceramic powders suggest that the price of the Si_3N_4 and SiC powders of engineering ceramic grade can be reduced down to 3–4 yen/gm. when production quantity amouts to 2,000 ton/year.

It is generally calculated that the sintered ceramic body can be manufactured at three to five times the cost of raw powder. According to this calculation, it is nearly impossible to manufacture the ceramic components of 6 yen/gm. from raw powders of 3–4 yen/gm.

An as-sintered Al_2O_3 ceramic body for spark plugs can be manufactured at a cost of

0.45–0.5 yen/gm. from 0.2 yen/gm. raw powder, low soda alumina. The ratio of the as-sintered ceramic body to raw powder is only 2.5 times in this case. About 400 million spark plugs per year are produced in Japan. The production process of Al_2O_3 ceramic body is as follows: grinding the low soda alumina, mixing the ground alumina powder and sintering additives, drying by spray dryer, forming by dry isostatic pressing, pre-sinter machining and sintering.[27] It should be noted that the processing cost for the as-sintered Al_2O_3 ceramic body is only 0.25–0.3 yen/gm. Therefore, it is reasonable to estimate that the PSZ ceramic body for the oxygen sensor can be manufactured not at a cost 3–5 times higher than its raw powder cost (30–50 yen/gm.), but at the cost of its raw powder cost plus 0.25–0.3 yen/gm. (10.25–10.3 yen/gm.), when the PSZ ceramic body is produced as much as the spark plug.

Non-oxide ceramics such as Si_3N_4 and SiC usually require higher sintering temperatures than oxide ceramics and an inert gas atmosphere in sintering. Therefore, the cost of facilities, especially furnaces, and expenses for the production of non-oxide ceramics are somewhat higher than those for oxide ceramics. When non-oxide ceramic components are manufactured in large quantities like spark plugs, however, the manufacturing cost excluding raw powder cost can be expected to reduce considerably. As a result, it is not necessarily impossible to manufacture the ceramic body at a cost of 6 yen/gm. from raw powder of 3–4 yen/gm. when the production quantity amounts to 2,000 ton/year.

Another item of concern is dimension accuracy. The Al_2O_3 part for spark plugs and the PSZ element for oxygen sensors, are not machined after sintering because the required accuracy is in the order of 100 μm. On the contrary, it is almost impossible to avoid post-sintering machining for the ceramic engine component because of the required accuracy of several μm.

It is calculated that as-sintered Si_3N_4 ceramic test pieces for bending strength can be manufactured at a cost less than 100 yen/pc, if as much as a hundred thousand pieces per month are produced by the process of dry pressing and pressureless sintering. However, the machining in conformity with JIS standard R-1601 may cost more than the manufacturing cost of the ceramic body. (Of course, there are actually few demands for the bending strength test piece. The price for such test pieces in Japan is about 2,000–10,000 yen/pc in compliance with the requested number.)

The following is another example of the machining price, and its dependence on the surface roughness:

    Dimension of test piece: $^\Phi 10 \times {}^L 100$ (mm)
    Machining section: all the cylindrical surface
    Ordered number: 30 pieces
    Requested surface roughness 2 μm: 4,500 yen/pc
                                0.8 μm: 11,000 yen/pc

It is feared that the cost of the post-sintering machining will be the greatest obstacle to realizing the ceramic engine. These numerical values suggest that more efforts must be devoted in the future to overcome the difficulty in machining Si_3N_4, SiC and other candidate ceramics.

There are three ways to decrease machining cost:

— Decrease the area, the accuracy and increase the roughness to be machined by the design.
— Decrease the quantity to be machined off by precise control of the manufacturing process.
— Develop new high speed and low cost machining technology.

Valve rocker arms with Si_3N_4 pad have recently been developed by NGK Insulators, Ltd. and Mitsubishi Metal Corp. in cooperation with Mitsubishi Motors Corp. It is noted that the sliding surface of the ceramic pad has been manufactured with such quality that it does not need post-sintering machining. The rocker arm was completed with no final machining for the ceramic part which was fixed in aluminium by diecasting, but the aluminium in the rocker arm was machined on the basis of the ceramic pad surface. This is a good example where an excellent ceramic manufacturing technology and a sophisticated design eliminated post-sintering machining.

Although there is such a good example, post-sintering machining is still one of the most expensive processes in the manufacturing of ceramic components for engines and the success of the ceramic engine, in a sense, depends on the excellent machining technology expected to be developed in the future.

13. 4 Conclusion

Major ceramic components applied to cars in the past decade are PSZ elements for the oxygen sensor and the monolith made of cordierite for catalytic converters. Although these ceramics were not necessarily new materials, their properties were significantly improved to meet the severe specifications requied.

Structural ceramics having excellent mechanical, thermal and chemical properties, such as Si_3N_4, SiC and PSZ, are candidate materials which can be used as components for the adiabatic engine and gas turbine engine. These ceramic applications will spread to other industries triggered by the development of automotive components.

There are three obstacles to overcome in order for ceramic components for engines to be used on a full scale: feasibility, reliability and cost. Of the three, "Cost" is the largest obstacle to overcome. Post-sintering machining is regarded as one of the most expensive processes in the manufacturing of ceramic components for a heat engine. Excellent machining technology, therefore, must be developed in the future.

In the maturing process of any technology, development generally comes in stages from basic research to applied research. On the other hand, when technology has the potential to create a field of industry, many development phases are performed in parallel. The development of ceramic materials and components for heat engines is now considered to be in such a parallel stage.

Acknowledgements

The author wishes to express his thanks to Dr. Y. Komatsu and Mr. M. Sasanouchi of Toyota Motor Corporation, Mr. M. Naito and Mr. N. Miwa of Nippondenso Co., Ltd, and Mr. T. Takeuchi, Dr. H. Doi and Mr. N. Kamiya of Toyota CRDL for their useful information; and to Dr. S. Kobayashi of Toyota CRDL for revision of the English text.

References

1) O. Kamigaito : J. Soc. Aut. Eng., **34,** 8 (1980) 811 [in Japanese].
2) Y. Shibata and T. Asano : Internal Combustion Engine, **21**, 258 (1982) 64 [in Japanese].
3) Y. Shibata, T. Asano and T. Hirano: Internal Combustion Engine, **21**, 259 (1982) 61 [in Japanese].
4) Y. Shibata and T. Asano : Internal Combustion Engine, **21**, 260 (1982) 69 [in Japanese].
5) Y. Shibata and T. Asano : Internal Combustion Engine, **21**, 261 (1982) 65 [in Japanese].
6) Y. Shibata and H. Banno : Internal Combustion Engine, **21**, 262 (1982) 46 [in Japanese].
7) Y. Shibata and H. Banno : Internal Combustion Engine, **21**, 263 (1982) 47 [in Japanese].
8) Y. Shibata and H. Banno : Internal Combustion Engine, **21**, 264 (1982) 57 [in Japanese].
9) Y. Shibata and H. Banno : Internal Combustion Engine, **21**, 268 (1982) 73 [in Japanese].
10) D. J. Godfrey : Science of Ceramics, Vol. 12, ed. by P. Vincentini, Grafiche Galeati, Italy (1984) 27.
11) R. A. Harmon and C. W. Breadsley : Mechanical Eng. (May 1984) 22.
12) R. A. Harmon and R. P. Larsen : Mechanical Eng. (Oct. 1984) 44.
13) R. Kamo : Ceramic Industry (July 1984) 26.
14) TOYOTA koho shiryo (TOYOTA Information), Jidosha to Zairyo (Automotive and Material) (April 1984) [in Japanese].
15) H. Kohl : Ber. Deut. Keram. Gesell., **13**, 2 (1932) 70 [in German].
16) I. M. Lachman and R. M. Lewis : U. S. Pat. 3885977 (May 27, 1975).
17) Japan Automobile Research Institute, Inc., Technical Note No. 7, A perspective survey on automotive technologies (April 1983) [in Japanese].
18) M. Ohashi: Eng. Mater., **31**, 12 (1983) 98 [in Japanese].
19) E. J. Horton and W. D. Compton : Science, 225 (1984) 587.
20) A. F. McLean : Ceramics for High Performance Applications, ed. by J. J. Burke, A. E. Gorum and R. N. Katz, Brook Hill Publishing Co., Chestnut Hill, MA (1974) 9.
21) Y. Hamano and M. Yamamoto : ASME Paper 84-GT-165 (1984).
22) S. Kato, K. Nishio, T. Sasaki and K. Kamino : Internal Combustion Engine, **23**, 300 (1984) 109 [in Japanese].
23) H. Kawamura : presented at the Ceramics Committee Meeting of Society of Automotive Engineers of Japan, Inc. (Nov. 30, 1984).
24) A. Kikuchi : Guide Book for High-Tech Materials Exhibition (1984) 6 [in Japanese].
25) Japan Automobile Manufacturers Association, Inc., Parts and Materials Committee, J. Iron and Steel Inst. Japan, **70**, 8 (1984) 218 [in Japanese].
26) H.K.Bowen:12th Automotive Materials Conference on "Ceramics in Engine" (March 1984).
27) S. Takagi : Fineceramics, **3**, 2 (1982) 64 [in apanese].

14. Pyroelectric Sensors of Lead Germanate Thick Films

Koichiro TAKAHASHI*

Abstract

Simple preparation techniques of pyroelectric sensors consisting of $Pb_5Ge_3O_{11}$ are reviewed and compared: (1) sintering technique, (2) glass-remelting-crystallization technique, (3) printing technique, and (4) rapid-quenching technique. Chemical and electric properties of the crystal are explained.

Keywords : pyroelectrics, lead germanate, film, ferroelectrics.

14.1 Introduction

Ferroelectric lead germanate, $Pb_5Ge_3O_{11}$, is of much interest in pyroelectricity. Many kinds of sensors are required in the mechatronics industry. Infrared sensors are especially necessary to detect relatively low temperature ranges of -30 to $300°C$ corresponding to 15-$5\mu m$ infrared. Pyroelectric sensors[1] are very convenient because they work well at room temperature and are relatively independent of temperature in detectivity.

These types of sensors have been mainly produced from single crystals. However, a disadvantage of this technique is its high cost due to cutting and polishing required, thus, inexpensive techniques have eagerly been sought. Simple preparation techniques for $Pb_5Ge_3O_{11}$(PGO) are explained in this review.

For PGO, together with TGS[2] and $LiTaO_3$[3] (commercially used as typical pyroelectric detector materials), various parameters for pyroelectric use are tabulated in Table 14.1[4] where T_c is the Curie temperature, ε the dielectric constant, p is the pyroelectric coefficient, and p/c_p' (c_p': volume specific heat) is the pyroelectric performance parameter. It is noted that PGO has a reasonably high pyroelectric coefficient, a relatively small dielectric constant and a high performance parameter, in comparison with other materials.

Lead germanate $Pb_5Ge_3O_{11}$(PGO) and its isomorphs[5-7] $Pb_5Ge_{3-x}Si_xO_{11}$(PGSO) with $0 \leq x \leq 1$ are ferroelectric at room temperature and optically active.[8,9]

Curie temperatures (T_c) range from 177°C for PGO to about $-30°C$ for PGSO with $x = 1.5$ near the limit of the solid solution series. The crystal structure of PGO[10] is hexagonal (P6) above T_c and trigonal (P3) below T_c.

---

* Muki-Zaishitsu Kenkyu-sho(National Institute for Research in Inorganic Materials), Namiki 1-1, Sakura-Mura, Niihari-Gun, Ibaraki 305, Japan.

Table 14.1 Pyroelectric sensor materials. [after K. Takahashi et al.[4]]

Materials	Shape	T_c(°C)	ε	p (C/cm^2°C) $\times 10^{-9}$	$p/\varepsilon c_p'$ (A·cm/W) $\times 10^{-10}$
TGS	sin. cryst.	49	35	40	4.6
PVF$_2$	film	120*	11	2.4-4.0	0.9-1.5
LiTaO$_3$	sin. cryst.	660	54	23	1.3
Sr$_{0.67}$Ba$_{0.33}$Nb$_2$O$_6$	sin. cryst.	67	1,700	110	0.31
PbTiO$_3$	sint. body	470	200	20	0.31
Pb(Zr$_{0.7}$Ti$_{0.3}$)O$_3$	sint. body	300	450	31	0.24
Pb$_5$Ge$_3$O$_{11}$	sin. cryst.	177	50	12	0.96
	sint. body	177	38	5	0.65

* Temperature at which polarization disruption occurs.

Jones et al.[11] have reported that the crystal is very suitable as a pyroelectric detector material, judging from the results of pyroelectric and ferroelectric measurements of both a single crystal and a sintered body.

Single crystals of PGO have been grown from melt using the Czochralski technique in <100> and <001> orientations.[12] The melting temperature of Pb$_5$Ge$_3$O$_{11}$ is about 740°C, and decreases only slightly with silica substitution.[13]

For pyroelectric device usage, the c-plane of the crystal is required because the c-axis is a polar axis. Several problems arise using the Czochralski technique[12] in preparation of the single crystal which is impossible to cut and polish due to its a-plane cleavages and resulting fragility.

Alternative techniques for preparation of a large and thin single crystal with the c-plane or ceramics consisting of oriented c-plane crystallites are strongly desired.

Eysel et al.[13] have suggested that small hexagonal platelets of PGO and PGSO can be produced by glass-remelting-crystallization. Takahashi et al.[14] have succeeded in obtaining <001> oriented thick-film single crystals with $0 \leq x \leq 1.5$ which are thin and wide enough to shape pyroelectric infrared detector chips[15].

Takahashi et al.[16] have also described a printing technique which is simple and inexpensive for preparing pyroelectric detector elements. Pinhole-free PGSO thin films (30-40 μm) with preferred orientations of a-plane and c-plane were successfully made in the range of $x = 0$-1.0. It is noteworthy that distinguished reorientation occurred even below melting temperatures.

According to Shimanuki et al.,[17] oriented pyroelectric glass-ceramics of the PbO-GeO$_2$ system were prepared using a rapid-quenching technique. A high degree of c-plane orientation of PGO crystallites was obtained in the thin samples.

14.2 Preparation and Pyroelectricity of Pb$_5$Ge$_3$O$_{11}$

14.2.1 Sintering Technique

For a PGO sintered body, both a preparation technique and electric properties are discussed.[18]

(a) Preparation of Sintered Body

Polycrystalline lead germanate silicate was obtained[14] by a solid state reaction at 550-600°C and 10-120 h using high-purity reagents.

$$5PbO + (3-x)GeO_2 + xSiO_2 \rightarrow Pb_5Ge_{3-x}O_{11} \ (0 \leq x \leq 1.5)$$

The obtained PGSO powder was formed into a disk at a pressure of 1 t/cm². The samples were sintered[18] at temperatures as low as about 20°C below the melting temperatures and silver paste was painted on both sides of the surface.

(b) Electric Properties of Sintered Body[18]

Temperature dependence of a specific dielectric constant (ε) is shown in Fig. 14. 1. With an increase in SiO_2 content, ε peak heights of PGSO sintered body decrease and become diffusive; moreover, the Curie temperature decreases. The ε-values was as large as 27-33 at room temperature, which is very low in comparison with those of PZT($\varepsilon = 450$).[4]

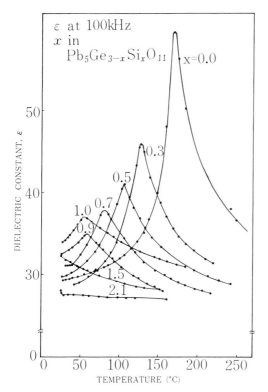

Fig. 14. 1 Specifitic dilectric constant (ε) of PGSO sintered bodies at 100 kHz. [after K. Takahashi et al.[18]]

Hasegawa et al.[19] have reported temperature dependence of ε in the samples crystallized from PGSO glasses in which the values at T_c are very low in comparison with Takahashi's results. It is surmised that their samples might retain the glassy portion which contributes to reduction of the dielectric constant. These results strongly suggest that this material has a highly structure-sensitive property.

Pyroelectric coefficients (p) of PGO sintered body and single crystal are illustrated in Fig. 14. 2.[20] The p-value of the former at room temperature is as large as 5×10^{-9}(C/cm²

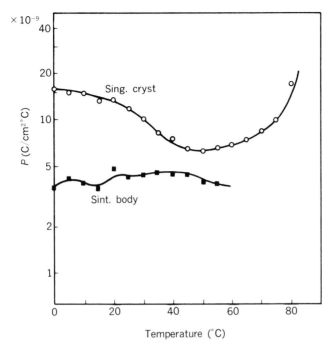

Fig. 14.2 Pyroelectric coefficient (p) of PGO single crystal and sintered body. [after K. Matsumoto et al.[20]]

°C), which is about half the value of the latter. We can regard the value of the sintered body as remarkably high, considering that electric-property-values of ceramics are generally one-tenth or one-hundredth as large as those of single crystals. Jones et al.[11] also recorded 5×10^{-9} (C/cm²°C) for a PGO sintered body.

The sintering technique is very simple and inexpensive; however, slicing and polishing are necessary for making a pyroelectric detector element. Simpler techniques are desired for practical use.

14.2.2 Glass-Remelting-Crystallization Technique

A very unique technique for preparation of PGO such as glass-remelting-crystallization[13,14] is introduced in this subsection.

(a) Preparation of PGO Single Crystal Films

Several different PGSO compositions were melted in gold crucibles heated to 800–820°C for 20 min. Glass is a more convenient starting material than polycrystalline powder because of its greater homogeneity and higher density. To form a glass, the melt was either quenched in water to give small particles or poured on an aluminum plate for pressing with a domestic iron to produce flat sheets. The presence of small crystallites in the light-yellow glasses caused no problems.

Recrystallization to the pyroelectric phase was carried out on a gold foil where the glass was remelted and then annealed at 600–720°C for 1–70 h. It is important that the remelted glass be not returned to room temperature before heat treatment. To obtain large

PGSO crystals, it is also important to flatten the melt surface by tilting the gold substrate, and to burst air bubbles in the melt.

Large hexagonal-shaped crystals up to 9 mm in the cross-section and less than 0.3 mm in thickness were precipitated by careful annealing. Micrographs of some of these light-yellow, transparent crystals are shown in Fig. 14. 3. The large lateral area and small thickness of the crystals is an ideal geometry for infrared detectors where large surface charge and small thermal mass are important. Back-reflection Laue photographs were taken to verify the crystalline phase and its orientation. The majority of crystallites are oriented with (001) parallel to the glass surface. This is very important because [001] is the direction of spontaneous polarization and maximum pyroelectric response.

In making glass-ceramics, the glass is generally recrystallized above the glass transition temperature after the melt has been poured out on a metal plate. Numerous nuclei appear in the glass during this type of heat treatment because the glass passes twice through the temperature range in which the nucleation rate is very high. Lead germanate silicate glasses are readily recrystallized to a polycrystalline ceramic in this way.

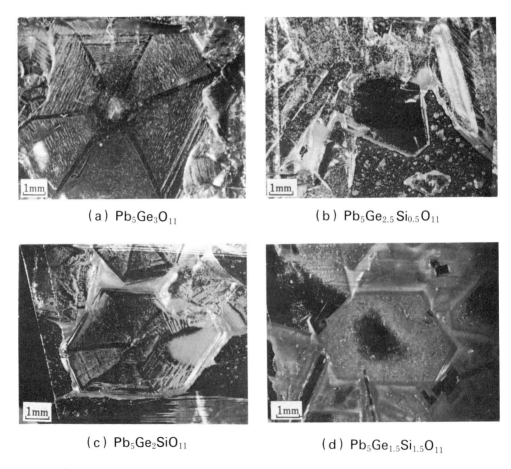

(a) $Pb_5Ge_3O_{11}$ (b) $Pb_5Ge_{2.5}Si_{0.5}O_{11}$

(c) $Pb_5Ge_2SiO_{11}$ (d) $Pb_5Ge_{1.5}Si_{1.5}O_{11}$

Fig. 14. 3 PGSO single crystals obtained by the glass-remelting-crystallization. [after K. Takahashi et al.[14)]]

A different procedure is required to produce large single crystals within the glass phase. The nucleation rate must be kept small while maintaining high crystal-growth rates. Takahashi et al.[21] have investigated surface nucleation rates and crystal growth conditions in sodium silicate glass for compositions near $Na_2O \cdot 2SiO_2$. Maximum crystal growth occurs at temperatures near 0.95 T_m, where T_m is the melting temperature in K. Maximum nucleation rates in sodium silicate occur at considerably lower temperatures near 0.46 T_m, and become very small above 0.54 T_m. Similar results were obtained[22] for $Li_2O \cdot 2SiO_2$ where the nucleation rate was a maximum at 0.58 T_m and negligibly small above 0.68 T_m.

It therefore appears that nucleation in viscous silicate systems is difficult for temperatures above 0.7 T_m. In principle no crystal nuclei appear above this temperature, but in practice contaminants always introduce the few nuclei, which produce large crystals. These nuclei grow to large sizes when annealed at temperatures near 0.95 T_m. The results for PGSO glasses are consistent with those for $Na_2Si_2O_5$ and $Li_2Si_2O_5$ glasses. When annealed at low temperatures, lead germanate silicate glass converts to a ceramic with many small crystallites, but when annealed at high temperatures (0.85 T_m to 0.96 T_m) large crystals are formed.

(b) Electric Properties of PGO Single Crystal Films

Prior to evaporation of Cr and Au on the as-grown surface without slicing and polishing, the plate-like crystals were cut as large as 4 × 4 mm² with a wire-saw. A front electrode was taken out of the Cr-Au evaporated surface and a back electrode from the gold foil.[15]

The electric factors of the PGO and PGSO samples prepared by the glass-remelting-crystallization are given in Table 14.2, where T_c(°C) is Curie temperature, D(%) is the dissipation factor at 10 kHz, P_r(C/cm²) is the remanent polarization and E_c(kV/cm) is the coercive field. The factors D, P_r, p and E_c are values at room temperature in the direction parallel to the c-axis. The temperature dependences of these values are illustrated in Figs. 14.4 and 14.5.

The above results suggest that the electric properties of the PGO and PGSO samples prepared by this technique are quite comparable to those prepared by the Czochralski technique.[23]

The frequency dependences of responsivities R_v(V/W) of the PGO and PGSO are

Table 14.2 Electric properties of PGSO single crystals prepared by the glass-remelting-crystallization and the Czochralski techniques. [after K. Takahashi et al.[15]]

Composition	Prep. tech.	T (°C)	ε R.T.	T_c	D (%) R.T. 10 kHz	P_r (C/cm²) R.T. ×10⁻⁶	p (C/cm²°C) R.T. ×10⁻⁹	E_c (V/cm) R.T. ×10³
$Pb_5Ge_3O_{11}$	Cz	177	41	3,000		4.8	9.5	16
	GRC	178	44	820	0.3	3.48	20*	11
$Pb_5Ge_2SiO_{11}$	Cz	60	150	900		1.7		4.8
	GRC	45	340	720	3	1.51	65*	6.4

* Estimated from Fig. 14.5

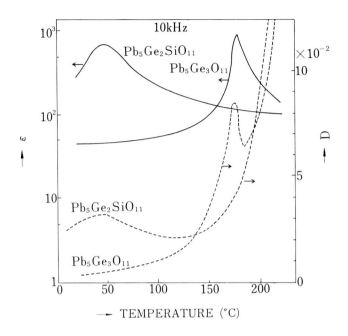

Fig. 14.4 Temperature dependence of dielectric constant (ε) and dissipation factor D (%) for PGSO by the glass-remelting-crystallization. [after K. Takahashi et al.[15])]

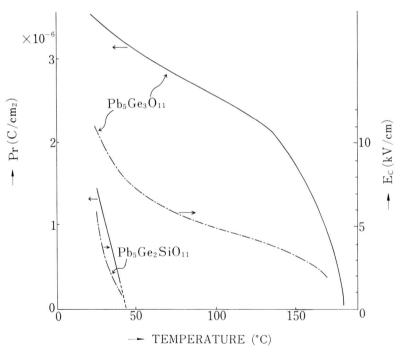

Fig. 14.5 Temperature dependence of remanent Polarization P_r(C/cm²) and coercive field E_c (kV/cm) for PGSO single crystals by the glass-remelting-crystallization.[after K. Takahashi et al.[15)]]

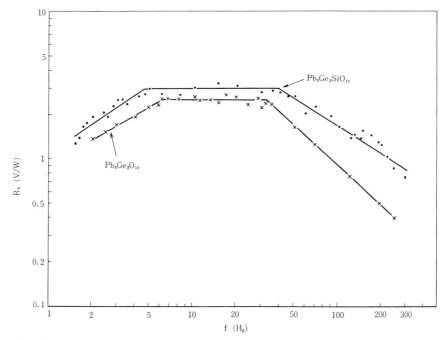

Fig. 14.6 Frequency dependence of pyroelectric responsivity R_v (V/W) for PGSO single crystals by the glass-remelting-crystallization. [after K. Takahashi et al.[15]]

shown in Fig. 14.6, whose curves are typical cases obeying Putley's equation.[2] It is evident that the pyroelectric performances of the PGO and PGSO are comparable to those of the commercial detectors such as TGS and LiTaO$_3$; furthermore, lead germanates, in particular, are superior to TGS in chemical stability.

14.2.3 Printing Technique

A printing technique[16,18] is an other simple technique for preparing PGSO thick films.

By adequately mixing pre-reacted PGSO powder and viscous organic agents (vehicles) with a pestle, paste was prepared, then printed using a screen on a gold foil. After drying at 110°C, the printed matter was fired in an electric furnace at 400°C for about 1 h to eliminate the vehicles. Finally the sample was heat-treated at 690–730°C for 0.5 h slightly below the melting temperatures. When pinholes were detected by observing the surface of the sintered film using a stereomicroscope, the paste was printed again on it, followed by firing.

The pinhole-free PGSO thick films (30–40 μm thick) were successfully made by the printing technique. Variations in grain texture of the sample with $x = 0.3$ during sintering process are shown in Fig. 14.7. At the lower temperatures, the shape of the grains was obscure, while it changed into a rectangle or hexagon, and grain growth was remarkable at the higher temperatures (the average grain size was about 15 μm at 710°C). Pinholes were found to disappear at the higher temperatures. With an increase in SiO$_2$ content, grain growth became remarkable at lower temperatures. This fact corresponds to a decrease in

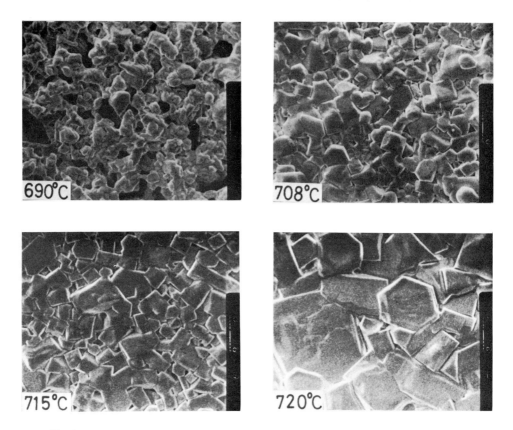

Fig. 14. 7 SEM photographs of the surfaces of PGSO thick films prepared by the printing technique, showing variation of grain growth and orientarion due to thermal history. [after K. Takahashi et al.[18)]]

melting temperature with an increase in SiO_2 content.

An X-ray diffraction pattern was recorded as shown in Fig. 14. 8 when X-ray was radiated vertically to the surface of the PGSO film. The intensities of the diffraction angles degree of orientation (F)[24)] reaches 7 to 39% in the c-plane (F_c) and 21 to 53% in the being attributed to a-plane or c-planes are remarkably strong as shown in Fig. 14. 8(b). The a-plane (F_a), which depend on x.[18)] It is found that the F-values of the a-plane are larger than those of the c-plane in almost the entire range of x. With an increase in x, the F-value has a tendency to decrease in the c-plane, but increase in the a-plane.

In the preparation of PGSO single crystals by the glass-remelting-crystallization technique,[14)] plate-like single crystals with c-plane orientation can be made by cooling the melts. It is of much interest that remarkable reorientation occurred even below the melting temperature in this synthesis. Judging from the fact that degree of orientation of a-plane and c-plane was distinguished, these two planes may be considered to be the most stable ones.[12)] The grains were oriented at random in the film just after printing. When the film was heated just below T_m, the constituent ions or atoms are thermally activated, so that mass transfer occurs from the part with high surface energy (e. g. edges of the grains and small particles) to the ones with low surface energy or from the surfaces with high energy to the ones with

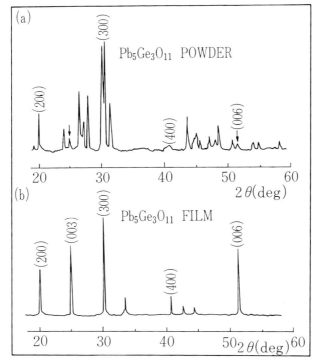

Fig. 14. 8 Comparison of X-ray diffraction patterns of PGO.
(a) the powder sample without any orientation, and (b) the thick film with preferred orientation prepared by the printing technique (heat-treated at 725°C for 1 h) [after K. Takahashi et al.[16] and K. Takahashi et al.[18]]

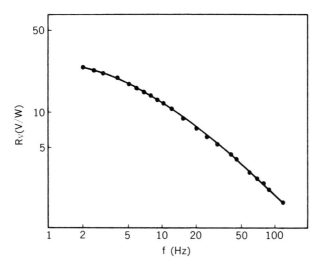

Fig. 14. 9 Frequency (f) dependence of pyroelectric responsivity (R_v) of $Pb_5Ge_{2.7}Si_{0.3}O_{11}$ thick film prepared by the printing technique. [after K. Takahashi et al.[18]]

low energy. Finally the ions or atoms may rearrange themselves into such a state that the planes with the lowest surface energy (e. g. a-planes or c-planes) appear on the surface.

Figure 14. 9[18)] shows frequency (f) dependence of pyroelectric responsivity (R_v) of the thick film with $x = 0.3$ prepared by the printing technique. This is characteristic of pyroelectric sensors, suggesting the high possibility for its use as a commercial IR detector.

14.2.4 Rapid-Quenching Technique

Shimanuki et al.[17)] have reported on the preparation of highly-oriented PGO films.

Glasses containing 62.0–63.5 mol% PbO in the PbO-GeO$_2$ system were prepared, where the stoichiometric composition of the Pb$_5$Ge$_3$O$_{11}$ crystal was 62.5 mol% PbO. To form a glass, the melt was rapidly cooled by four quenching techniques; twin roller, single roller, dipping, and roller-plate techniques (Fig. 14. 10) designed by Shimanuki et al. In preparation of glass-ceramics, crystallization was carried out by placing as-quenched glass specimens on a gold-evaporated alumina substrate, followed by annealing isothermally at 600–720°C for 1–24 h at a heating rate of 200°C/h.

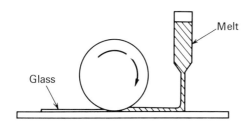

Fig. 14. 10 Schematic view of the roller-plate technique. [after S. Shimanuki et al.[17)]]

Of the four techniques, the roller-plate one was found to be the most suitable for preparing lead germanate glasses with large area and small thickness of 0.1–1 mm. Lead germanate glasses prepared by this technique were transparent and yellowish.

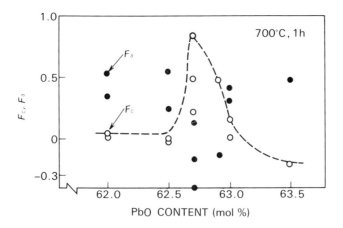

Fig. 14. 11 Compositional dependence of degree of orientation, F_c and F_a for the PbO-GeO$_2$ system obtained by the roller-plate technique. [after S. Shimanuki et al.[17)]]

By this technique, highly-oriented films of PGO were obtained; particularly the degree of c-plane and a-plane orientation (F_c and F_a)[24] was remarkably large. For the glass-ceramics crystallized at 700°C for 1 h, the values of F_c and F_a are plotted against composition in Fig. 14.11. It is noted that the glass-ceramic with 62.7 mol% PbO showed the maximum value of F_c.

On the other hand, annealing temperature dependence of F_c and F_a is illustrated in Fig. 14.12 for the glass-ceramics with the stoichiometric and the PbO-rich compositions. The samples with the stoichiometric composition has a tendency toward low F_c and high F_a in the range of 600–710°C in annealing temperature, while the PbO-rich composition (62.7 mol%) reaches a very high F_c of 0.84 at 700°C.

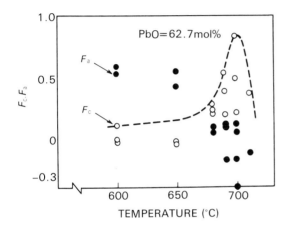

Fig. 14.12 Annealing temperature dependence of F_c and F_a for the PbO-GeO$_2$ system by the roller-plate technique. [after S. Shimanuki et al.[17]]

Shimanuki et al.[17] suggested that such high orientation is due to surface nucleation and the anisotropic grain growth of $Pb_5Ge_3O_{11}$ crystallites. For the c-plane orientation, it is explained that the grain growth in the a-axis direction of PGO becomes dominant due to the liquid phase which appears in the PbO-rich composition, and therefore, large c-plane PGO crystallites are formed parallel to the thin sample by the discontinuous grain growth.

Glass et al.[25] reported that the p-value of the PGO glass-ceramic was as large as 5.4×10^{-9} (C/cm^2·°C) at room temperature, which is comparable to the value of Matsumoto's for the sintered body.[20]

The oriented films prepared by Shimanuki et al. are expected to give much higher pyroelectric coefficients, and furthermore, to be used as a pyroelectric detector element.

14.3 Conclusion

In the review of pyroelectric detectors of ferroelectric lead germanates, four simple preparation techniques were compared: sintering, glass-remelting-crystallization, printing, and rapid-quenching techniques. These techniques share the following merits:
(1) The techniques are simple and require no sophisticated large-scale preparation instruments. Troublesome slicing and polishing processings are eliminated.
(2) Crystal size is sufficiently large, yet thin enough to be directly used as a detector element except in the sintering technique.

(3) The polar axis is almost perpendicular to the crystal surface except in the sintering technique.

(4) Electric properties of the obtained samples are adequate for a pyroelectric detector.

Acknowledgements

The author wishes to express his gratitude for valuable discussions to the following persons: Prof. Leslie E. Cross and Prof. Robert E. Newnham(Mater. Res. Lab., Pennsilvenia State University), Mr. Keiji Takamatsu and Dr. Kazutoshi Matsumoto (Central Res. Lab., Sumitomo Metal Mining Co. Ltd.), Mr. Senji Shimanuki (Res. & Dev. Center, Toshiba Corporation), and Dr. Kazuyuki Kakegawa (Faculty of Eng., Chiba University).

References

1) E. Yamaka : "Pyroelectric phenomenum of crystals and their applications to infrared detection," Oyo Buturi, **43**, 2 (1974) 153[in Japanese].
2) E. H. Putley : "The pyroelectric detector," Semiconductors and Semimetals, Vol. 5, ed. by R. K. Williams and A. C. Beer, Academic Press, N. Y. (1970) p. 259.
3) C. B. Roundy and R. L. Byer : "Sensitive $LiTaO_3$ pyroelectric detector," J. Appl. Phys., **44**, 2 (1973) 929.
4) K. Takahashi, K. Takamatsu and F. Nakayama : "Development of polycrystalline infrared detector elements and their applications," Keiso, **27**, 7 (1984) 26[in Japanese].
5) H. Iwasaki, K. Sugii, T. Yamada and N. Niizeki : "$5PbO \cdot 3GeO_2$ crystal: a new ferroelectric," Appl. Phys. Lett., **18** (1971) 444.
6) S. Nanamatsu, H. Sugiyama, K. Doi and Y. Kondo : "Ferroelectricity in $Pb_5Ge_3O_{11}$," J. Phys. Soc. Jpn., **31** (1971) 616.
7) J. Dougherty, E. Sawaguchi and L. E. Cross : "Ferroelectric optical rotation domains in single-crystal $Pb_5Ge_3O_{11}$," Appl. Phys. Lett., **20** (1972) 364.
8) H. Iwasaki, K. Sugii, N. Niizeki and H. Toyoda : "Switching of optical rotatory power in ferroelectric $5PbO \cdot 3GeO_3$ single crystal," Ferroelectrics, **3**, 2-4 (1972) 157.
9) R. E. Newnham and L. E. Cross : "Ambidextrous crystals," Endeavour, **118** (1974) 18.
10) R. E. Newnham, R. W. Wolfe and C. N. W. Darlington : "Prototype structure of $Pb_5Ge_3O_{11}$," J. Solid State Chem., **6** (1973) 378.
11) G. R. Jones, N. Shaw and A. W. Verre: "Pyroelectric properties of lead germanate," Electron. Lett., **8**, 14 (1972) 345.
12) K. Sugii, H. Iwasaki and S. Miyazawa : "Crystal growth and some properties of $5PbO \cdot 3GeO_2$ single crystals," Mater. Res. Bull., **6** (1971) 503.
13) W. Eysel, R. W. Wolfe and R. E. Newnham : "$Pb_5(Ge,Si)_3O_{11}$ ferroelectrics," J. Am. Ceram. Soc., **56**, 4 (1973) 185.
14) K. Takahashi, L. E. Cross and R. E. Newnham : "Glass-recrystallization of ferroelectric $Pb_5Ge_3O_{11}$," Mater. Res. Bull., **10**, 7 (1975) 599.
15) K. Takahashi, L. H. Hardy, R. E. Newnham and L. E. Cross : "Pyroeffect in $Pb_5Ge_3O_{11}$ and $Pb_5Ge_2SiO_{11}$ monocrystals prepared by glass-recrystallization," 1979 Proc. 2nd Meeting on Ferroelectric Materials and Their Applications (1979) 257.
16) K. Takahashi, S. Shirasaki, K. Takamatsu, N. Kobayashi, Y. Mitarai and K. Kakegawa : "Oriented $Pb_5Ge_{3-x}Si_xO_{11}$ thick films prepared by the printing technique," J. Mater. Sci. Lett., **3**, 3 (1984) 239.
17) S. Shimanuki, S. Hashimoto and K. Inomata : "Oriented grain growth from lead germanate glasses," Ferroelectrics, **51**, Part 2 (1983) 53.
18) K. Takahashi, S. Shirasaki, K. Takamatsu, N. Kobayashi, Y. Mitarai and K. Kakegawa : "Pyroelectricity of preferably-oriented $Pb_5Ge_{3-x}Si_xO_{11}$ thick films prepared by the printing technique," Jpn. J. Appl. Phys., **22**, Suppl. 22-2 (1983) 73.

19) H. Hasegawa, M. Shimada and M. Koizumi: to be published.
20) K. Matsumoto, N. Kobayashi, K. Takada, K. Takamatsu, H. Ichimura and K. Takahashi: "Dielectric properties of ceramic lead germanate derivatives," Jpn. J. Appl. Phys., **24**, Suppl.24-2 (1985) 466
21) K. Takahashi and T. Sakaino : "Influence of atmospheres on the surface crystallization of $Na_2O \cdot 2SiO_2$ glass," Bull. Tokyo Inst. Technol., 104 (1971) 1.
22) M. Ito, T. Sakaino and T. Moriya : "Study on the process of crystallization of the $Li_2O \cdot 2SiO_2$ glass, I rates of crystal growth and nucleation," Bull. Tokyo Inst. Technol., 88 (1968) 127.
23) H. Iwasaki, S. Miyazawa, H. Koizumi, K. Sugii and N. Niizeki : "Ferroelectric and optical properties of $Pb_5Ge_3O_{11}$ and its isomorphous compound $Pb_5Ge_2SiO_{11}$," J. Appl. Phys., **43**, 12 (1972) 4907.
24) F. K. Lotgerling : "Topotactical reactions with ferrimagnetic oxides having hexagonal crystal structures-I," J. Inorg. Nucl. Chem., **9** (1959) 113.
25) A. M. Glass, K. Nassau and J. W. Shiever : "Evolution of ferroelectricity in ultra-grained $Pb_5Ge_3O_{11}$ crystallized from the glass," J. Appl. Phys., **48**, 12 (1977) 5213.

15. Ferrite Materials

Takeshi NOMURA*, Katsunobu OKUTANI* and Tatsushiro OCHIAI*

Abstract

Recent developments in the field of soft and hard ferrite materials are reviewed with reference to production conditions. The electromagnetic properties of ferrites, soft or hard, depend on the production process and microstructures. Precise and well-planned control of the production process is a key to improve quality, as well as to reduce the cost of ferrites.

Keywords : ferrite, spinel, hexagonal, permeability, single crystal.

15.1 Introduction

Ferrite materials have been used extensively in various coils, as well as in permanent magnets. Ferrites, having many advantages over other magnetic materials, such as alloys, have fulfilled a major need of the rapidly growing electronics industry; thus, the ferrite industry has grown to a considerable size. Generally, ferrites used in coils are called soft ferrites, while those used as permanent magnets are called hard ferrites. Almost all the soft ferrites have a spinel structure and high permeability. In the other hand, hard ferrites have a magnetoplumbite structure, high coercive force, and high residual flux density. Representative soft ferrites are MnZn-ferrites and NiZn-ferrites, and typical hard ferrites are Sr-ferrites and Ba-ferrites. Soft magnetic materials include metallic magnetic materials and ferrites. Metallic magnetic materials such as permalloys, Fe-Si alloys and amorphous alloys have advantages such as high permeability at low frequencies, high saturation magnetization and high Curie temperature. However, major disadvantages of these materials are low permeability at high frequencies and low resistivity. At high frequencies, contrarily, ferrites are superior to metallic magnetic materials. Ferrites have excellent low loss characteristics because of their high resistivity, in spite of their low saturation magnetization, low Curie temperature and low thermal conductivity. Two representative soft ferrites, MnZn ferrites and NiZn ferrites, are widely used in various applications such as fly-back transformers for TV sets, telecommunication applications, and recording heads. In such applications, MnZn ferrites have major advantages over ferrites with high saturation magnetization and high initial permeability. Thus, demands for MnZn ferrites are increasing, especially in power applications, despite their relatively low resistivity. On the other hand, NiZn ferrites are used in high-frequency applications because of their higher resistivity.

Two other types of hard magnetic materials are metallic and ferrite magnets. Metallic

---

* Ferrite Research and Development Laboratory, TDK Corporation, 2-15-7, Higashi-Ohwada, Ichikawa, Chiba 272-01, Japan.

magnets such as Al-Ni-Co-Fe alloys and rare-earth alloys have high residual flux density and high maximum energy product. On the other hand, ferrite magnets have advantages over metallic magnets such as low cost and high coercive force. Much effort have been devoted to develop high performance ferrites. In both soft and hard ferrites, electromagnetic properties are strongly influenced by the composition, the production process and ferrite microstructure. Ferrite production has grown remarkably in the past years to keep up with the boom in electronics markets: MnZn ferrites for switching power supplies and magnetic recording heads, and Sr ferrites for acoustic devices and motor magnets. The aim of the present paper is to survey recent developments in the field of ferrites for switching power supply, magnetic recording heads and permanent magnets. However, some types of ferrite materials for special use, such as magnetic recording media, magnetic fluid, and carriers for electrophotocopy will be omitted.

15.2 MnZn Ferrites

MnZn ferrites are generally classified into three groups, high initial permeability materials for wide band and pulse transformers, low loss materials for inductors and telecommunication use, and high saturation flux-density materials for power transformers and magnetic recording heads. Ferrite performance is not determined by the high value of initial permeability alone; a low loss value, represented by quality factors, relative loss factors and power loss, is also important. Moreover, high saturation flux density, high fired density, and frequency characteristics are necessary considerations for power transformers and magnetic recording heads. In many cases, these requirements are not satisfied at the same time; so that the optimum material is to be selected for such usage.

15.2.1 Initial Permeability

The initial permeability values (μ_i) depend on the mobility of the Bloch's domain wall.[1] To obtain high μ_i it is important to lower the anisotropy and to increase the saturation magnetization. Three important factors govern the value of μ_i; composition, additives and microstructure. Generally, the μ_i-temperature curve shows two major peaks. One is a maximum just below the Curie temperature caused by the abrupt decrease of anisotropy. The other is a secondary peak maximum near room temperature. Thus, there are two ways to achieve high μ_i. One is to adjust the composition to push down the Curie temperature to the working temperature. However, to do so is not recommended because of the instability of μ_i with temperature fluctuation. Therefore, commercial high μ_i materials are obtained by adjusting the composition to get the secondary peak maximum at the working temperature. Another way is to compensate the negative magnetocrystalline anisotropy (K_1) of the ferrite by substituting suitable ions which have positive K_1 such as Fe^{2+} and Co^{2+}.[2] Compensation by Fe^{2+} is more usual in MnZn ferrites. Fe^{2+} ions are introduced either by introducing an excessive amount of Fe_2O_3 into the stoichiometric composition or by substituting ions with higher valencies such as Ti^{4+} or Sn^{4+}.[3] Even if $K_1 = 0$, other anisotropy such as K_2[4] or magnetoelastic anisotropy does not disappear. Thus, besides $K_1 = 0$, λ_s(magnetostriction constant) $\doteqdot 0$ is another important factor to obtain higher μ_i.[5] The effect of composition on μ_i is shown in Fig. 15.1. The magnetostriction may cause and anisotropy introduced mechanically. Many studies have been done from this view point.[6-8] In almost all cases, a strain induced mechanically or chemically (compositional fluctuation in the sintered body or a microstrain caused by a grain boundary) decreases μ_i. μ_i may be responsible for the mobility of the domain wall. Thus, it is thought that inclusions such as pores or nonmagnetic impurities, which may pin the domain wall and causes gaps, should be avoided. Raw materials of MnZn ferrites for high μ_i materials should be of high purity, and the high fired density must be ensured. It is well known that μ_i increases linearly with grain size,[9-10] and μ_i up to 40,000 is achieved with grains of 80 μm. The MnZn ferrite production process, including the firing process, should be selected along this line. Ambient atmosphere is important[11-13] because the oxygen content of MnZn ferrite has a strong effect on μ_i, as is shown in Fig. 15. 2.

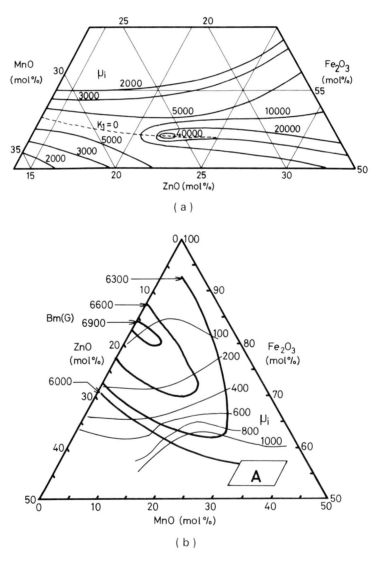

Fig. 15. 1 Magnetic properties of the MnZn ferrites represented in the miscibility diagram. [(a) : after E. Röss[9]]

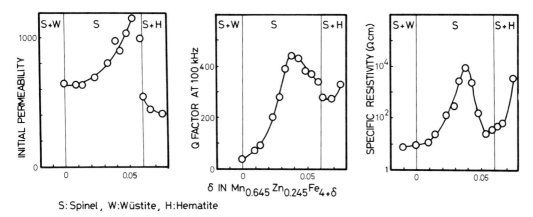

Fig. 15. 2 Effect of oxygen content on the magnetic properties of the MnZn ferrite. [after A. Morita and A. Okamoto[11]]

15.2.2 Loss Characteristics

Low loss MnZn ferrites should have uniform fine grain size and high fired density. Thus far, these requirements have been attained mainly by the selection of additives such as TiO_2 and SnO_2 which lower the firing temperature without accelerating grain growth.[3] Moreover, in the conventional manufacturing process for low loss MnZn ferrites, ideal magnetic properties, such as low relative loss factors and low disaccommodation factors can be obtained by controlling various parameters to attain the optimum oxygen content as is shown in Fig. 15. 2.[11] Eddy current loss is predominant over other losses such as hysteresis loss or residual loss at high frequencies. The eddy current loss may be reduced effectively by a uniform and fine-grained microstructure, as well as by a grain boundary with high electrical resistivity. TiO_2 and SnO_2 are known to increase the electrical resistivity not only in the low loss materials[14] but also in high flux density materials.[15] High electrical resistivity is required to obtain high permeability at higher frequencies. The firing temperature of the power ferrites is higher than that of low loss materials to attain high fired densities.

Thus far, power loss characteristics of ferrites have been improved mainly by adequate additives. Additives and impurities responsible for the grain boundary chemistry have a marked effect on the microstructure. Thus, the selection of adequate additives is very important. To obtain a sintered body with uniformly sized fine grains, grain growth should be suppressed especially in the initial stage of sintering. This is done by controlling oxygen discharge, for which the heating rate and ambient atmosphere are responsible as is shown in Fig. 15. 3.[13,16] The slow heating or vacuum firing in the initial stage of sintering, for instance, is effective. In the initial stage of sintering, the heating rate should be small enough to ensure the uniform grain growth in the subsequent firing. Adequate control of oxygen partial pressure is also important to prevent anomalous pore growth caused by the formation of high-concentration cation vacancies at high oxygen partial pressure.[17] Figure 15. 4 illustrates the results of Auger analysis of the fracture surface of three kinds of MnZn

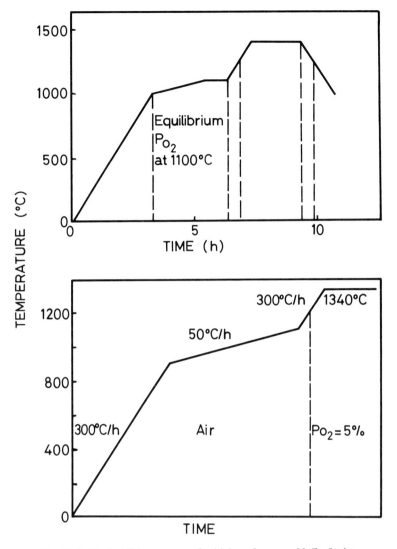

Fig. 15. 3 Typical firing program for high performance MnZn ferrite.

ferrites.[16] It is well known that adequate control of SiO_2 and CaO contents is important to obtain low loss characteristics.[14,18] In Fig. 15. 4, SiO_2 and CaO contents at the grain boundary are indicated. A commercial ferrite of high μ_i is composed of thin grain boundaries. On the other hand, in the case of a conventional high B_m material for power transformers, grain boundaries are usually thick for reducing the eddy current loss. Higher concentrations of impurities such as SiO_2 and CaO at the thin grain boundaries are desirable for power ferrites, as is seen in the modified material. Adequate selection of additives, firing program and production process of raw material powder would achieve such desirable microstructures reproducibly.

The relationship between relative loss factor ($\tan\delta/\mu_i$) and power loss is shown in Fig. 15. 5.[16] It is shown that low power loss is achieved when the relative loss fator is low. Relative loss factor shows the loss under weak magnetic field. On the other hand, power loss

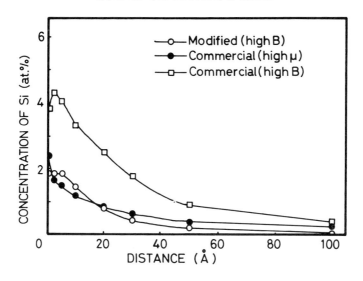

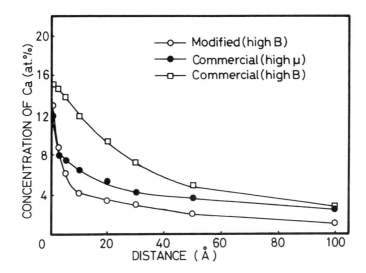

Fig. 15. 4 Auger analysis of the fracture surface of three kinds of MnZn ferrites.

shows the loss under strong magnetic field. This figure means that relative loss factor and power loss can be improved in the same manner.

15.2.3 Saturation Flux Density

Saturation flux density of MnZn ferrites is a strong function of fired density, temperature and composition, as is shown in Fig. 15. 1 and the following empirical equation:[19]

$$B_{ST} = (\rho/\rho_t)[1 - (T/T_c)^n]B_{S0} \tag{15.1}$$

where, B_{ST}: saturation flux density at $T(K)$

B_{S0}: saturation flux density at 0 K

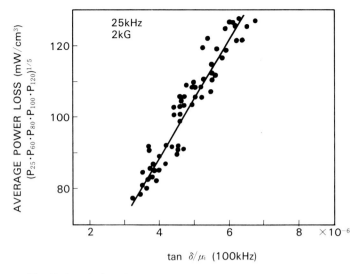

Fig. 15.5 Relationship between relative loss factor and power loss.

ρ: fired density

ρ_t: theoretical density

T: temperature

T_c: Curie temperature

n: constant ($\fallingdotseq 2$).

Generally, working temperature ranges from room temperature to about 100°C. Then, high fired density and high B_{S0} are important to obtain high B_{ST} at the working temperature. B_{S0} is a strong function of composition and cation distribution between A and B sites in the spinel structure, which is governed by firing conditions. Maximum value of saturation flux density attained in the Mn-Zn-Fe-O system is 6,900 G with the single crystal.[20]

As is shown in Fig. 15.1, the higher the Fe_2O_3 content, the higher the saturation flux density. Other magnetic properties such as μ_i, residual flux density and coercive force, however, become unfavorable with increasing Fe_2O_3 content. Power ferrites and ferrites for magnetic recording heads are required to have high saturation flux density, high initial and/or amplitude permeability, high Curie temperature and secondary peak maximum located at about 100°C. Thus, the composition A should be selected for these applications. In the case of ferrites for magnetic recording heads, high fired density is most important, because the pores in the fired body may cause an unexpected gap in the recording head and is unfavorable for wear resistance during head-tape driving. In MnZn ferrites, high fired densities are difficult to obtain in atmospheric firing. Vacuum sintering and pressure sintering techniques have been developed. Using the vacuum sintering technique, over 98 % of the theoretical density can be obtained. Pressure sintering techniques are hot pressing and hot isostatic pressing (HIP) which can give high fired density over 99.9% of the theoretical density. Generally, HIP is operated at about 1,200°C and about 1,000 atm. HIP needs a pre-sintering step, usually done by vacuum sintering. These special techniques result in high cost; thus, the use is limited to high performance ferrites for magnetic recording head.

15.2.4 Production Process

It is well known that the raw material has a marked effect on the electromagnetic properties of final products. Many studies have been done from this point of view.[21-24] The effect of the source material on the final product, however, has not been fully understood. The effect of SiO_2, CaO, TiO_2, SnO_2 and MoO_3 contents on the quality of MnZn ferrites is well known.[14,15,25] In contrast, the effect of other impurities or additives has not been investigated, or rather, is top secret for ferrite manufacturers.

It is important to note that the adequate selection of raw materials and processing parameters is a key to the control of microstructure and electromagnetic properties. In other words, raw materials of high purity and a well-planned process which makes the best use of raw material charateristics are important to make high performance ferrites. Much effort has been devoted to develop new ferrite manufacturing processes for both high quality and low cost. Unfortunately, no promising process seems to have been established thus far. There are three major methods for obtaining ferrite and/or ferrite raw material powders ready to press. The most conventional one is called a dry method, consisting mixing, calcination, comminution, etc. Because of its of low cost, this method has an advantage over a wet method such as coprecipitation. Disadvantages of this method include difficulties in obtaining fine and sharply distributed particles, difficulties in avoiding high level contamination, and difficulties in controlling powder characteristics. The second method is a wet method such as coprecipitation, sol-gel, and freeze-drying methods. These methods supply ferrite powders of high purity, fine and sharp distribution of primary particle size and high physical and chemical homogeneity. They are, however, costly because more processing steps are required than other methods. Thus, only the coprecipitation method is adopted for practical production and its use is limited to special use in high performance ferrite cores. The last method is a spray method, such as spray firing[28] and spray roasting.[29,30] The calcination process is an important process in the fabrication of soft ferrites for controlling the powder properties thought to be responsible for the electromagnetic properties of final products. In the spray firing method, the dry premixed starting materials (oxides, carbonates and/or hydroxides) are suspended into deionized water by a wet mill. Then, the slurry is sprayed into the spray calciner of the Ruthner unit. In this technique, powder properties can be easily controlled by the process parameters during calcination. In the spray roasting method, on the other hand, iron chloride solutions (pickled liquor) in combination with suitable metal chloride solutions are thermally decomposed in a spray roaster to yield ferrite powders of desired compositions. This method has a major advantage in that it skips the calcination process. These processes are compared in Table 15.1. A distinct feature of the spray roasting process is in the powder processing. The performance of final products is strongly influenced by the raw materials, because particle size and distribution, chemical properties and other powder parameters of calcined materials reflect raw material characteristics. In the spray roasting and coprecipitation methods, however, size and distribution of primary particles are controlled mainly by the spray roasting conditions and precipitation conditions, respectively, and not by the starting materials. In short, powder characteristics are rather easy to control in spray roasting and coprecipitation methods. The conventional

Table 15.1 Comparison of conventional, coprecipitation and spray roasting processes.

Criteria	Conventional Process	Coprecipitation Process	Spray Roasting Process
Process	break down	build up	build up
Processing Steps	standard	more	less
Calcination	necessary	necessary	unnecessary
Comminution	long time	short time	short time
Cost Performance	good	not good	?
Powder Porperties			
Primary Particle Size	0.8 - 3.0 μm	0.05 - 0.5 μm	0.4 - 1.0 μm
Size Distribution	wide	narrow	narrow
Homogeneity	not good	good	good
Purity	low	high	high
Core properties	good	excellent	excellent

production process contains a long-time comminution step after calcination, such that a significant amount of impurities such as SiO_2 is introduced from the mill. In the spray roasting method, on the other hand, short-time comminution prevents the materials from being contaminated by the milling system. The number of processing steps in the conventional production method is more than that in the spray roasting method. Thus, the spray roasting method is much easier to control than other methods. This method will be a key to remarkable improvements in quality as well as the decrease in MnZn ferrite cost.

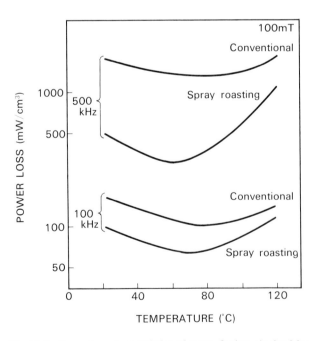

Fig. 15.6 Power loss characteristics of power ferrites obtained by conventional and spray roasting process.

Power loss characteristics of the improved power ferrite obtained by spray roasting process are shown in Fig. 15. 6. Improved material shows much lower power loss at higher frequency (500 kHz). This is because the material obtained by spray roasting is of high purity and uniformly sized fine grains.

15.2.5 Single Crystals of MnZn Ferrites

Single Crystals of MnZn ferrites are used as video tape recording head materials, and demand for single crystals of MnZn ferrites is rapidly increasing in keeping with the growing market of video tape recorders (VTR). Advantages of single crystals for VTR head materials are good wear-resistance and little chipping during processing into the head component. In using single crystals of MnZn ferrites for VTR head materials, it is important to have high μ_i at working frequencies, high saturation flux density, and high stability against operating conditions. There are many kinds of production methods of the single crystal of MnZn ferrite, but almost all of the single crystals of MnZn ferrites are manufactured using the Bridgeman technique. In this method MnZn ferrites are melted in a Pt crucible at about 1,700°C, and then cooled slowly to grow a single crystal from the melt on the seed crystal. The advantage of this method is the high productivity of big single crystals up to 3 in ϕ. In this method, however, the composition of the crystal is known to vary in the growth direction due to MnO segregation and ZnO evaporation from the surface of the melt during growth. Furthermore, Fe_2O_3 dissociates at high temperatures to form Fe^{2+}, and the melt is contaminated by Pt particles from the crucible. Many studies have been done[31-33] in order to improve the compositional uniformity, for example, by adding a starting material to the melt during crystallization. Recently, two new methods for the production of MnZn ferrite single-crystal have been developed. One is a travelling solvent zone melting method,[34] and the other is a solid-state reaction method.[35] In the travelling solvent zone melting method, the solvent is inserted between the single crystal and the starting material in the Pt crucible. Since the crystal grows while the starting material is being supplied to the solvent, the solute composition in the solvent is kept constant during growth, resulting in a uniform composition.

In the solid-state reaction method, a polycrystal is brought into contact with a seed crystal with or without HNO_3, then heated at temperatures at which discontinuous grain growth does not occur. The principle of this method is as follows: The driving force of growth of single crystal is proportional to the boundary curvature between the single crystal and polycrystal. This also holds true in the case of grain growth. Thus, the fine grained polycrystal is desirable. Using the polycrystal of uniform microstructure and homogeneous composition, a single crystal of highly uniform composition can be obtained. However, a disadvantage of this method is that some inclusions, such as pores in the polycrystal, remain in the single crystal; thus, a high quality polycrystal should be used. This method is one of the most promising ways to obtain MnZn ferrite single-crystal at low cost.

15. 3 Hard Ferrites

A hard ferrite has a hexagonal crystal structure with the uni-axial c-axis being the one along which the magnetization prefers to align. Technical requirements for ferrite magnets are high residual flux density and high coercive force. Representative hard ferrites are $BaO6Fe_2O_3$ and $SrO6Fe_2O_3$. $BaO6Fe_2O_3$ is lower in cost than $SrO6Fe_2O_3$. The energy product of $BaO6Fe_2O_3$ is, however, smaller than that of $SrO6Fe_2O_3$. Ferrite magnets are required to have low cost, high residual flux density and high coercive force. Generally, a hard ferrite with high residual flux density shows a low value of coercive force, while that with high coercive force shows a low value of residual flux density. Many studies have been done to attain high residual flux density and high coercive force at the same time.

15.3.1 Residual Flux Density

Residual flux density of hard ferrites is a function of saturation flux density, degree of orientation, and fired density, as expressed by the following equation:

$$I_r = k \cdot I_s \cdot n_c \cdot (\rho/\rho_0) \tag{15.2}$$

where, I_r: residual magnetization

I_s: saturation magnetization

n_c: degree of orientation

ρ: fired density

ρ_0: theoretical density

k: constant.

It should be noted that I_s, n_c and ρ must be high for high residual flux density (B_r). Present-day, M-type ferrites such as $SrFe_{12}^{3+}O_{19}$ of which I_s is 72 emu/g, and $BaFe_{12}^{3+}O_{19}$ are being manufactured. In the manufacturing process, a molar ratio of Fe_2O_3 to BaO, or Fe_2O_3 to SrO is not chosen as 6 but 5.5-5.9. These values are chosen for the following reasons: Excess Fe_2O_3 over the molar ratio of 6 will precipitate as a non-magnetic phase, and a considerable amount of Fe_2O_3 will enter the material as contamination during comminution. Moreover, it is reported[36] that the new complex ferrite, $Ba_3Fe_4^{2+}Fe_{28}^{3+}O_{49}$, exists in the $BaO5.5Fe_2O_3$ magnets. $Ba_3Fe_4^{2+}Fe_{28}^{3+}O_{49}$, was found to have the following characteristics at room temperature: Saturation magnetization $4\pi I_s = 5,000$ G and anisotropy field $H_A = 19.3$ kOe. Furthermore, intermediate phases enriched in SrO or BaO are reported to accelerate densification during firing.[37,38] In addition to the M-type ferrite, there are ferrites of other types in the $Ba(Sr)O-Fe_2O_3-MO$ system as shown in Fig. 15.7. W-type ferrite, $SrFe^{2+}Fe_{16}^{3+}O_{27}$, which is thought to have higher saturation magnetization (79 emu/g), is noteworthy for high B_r material. Since F. K. Lotgering et al.[39] had obtained pure W-type material by careful control of firing, many studies have been done on W-type ferrite from the viewpoint of substitution of Fe^{2+} by other divalent metallic ions such as Ni^{2+}, Co^{2+} and Zn^{2+} [40-42]

It should be noted that the degree of orientation is of prime importance for attaining high B_r. The production process of hard ferrite materials is illustrated in Fig. 15.8. Dry and wet pressing, and isotropic or anisotropic processes are chosen from the viewpoint of cost performance. High performance ferrite magnets are all manufactured by wet pressing in a magnetic field. In the present situation, the degree of orientation by pressing in a magnetic field is usually 80-90 %. To obtain a highly oriented pressed-body, pressing techniques, slurry conditions and powder properties such as particle shape, size and distribution are

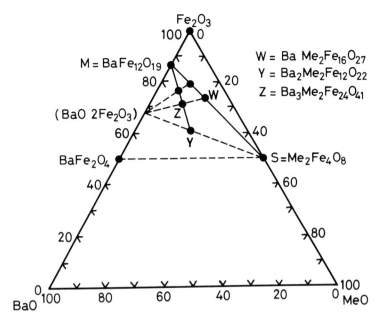

Fig. 15.7 Composition diagram for ferromagnetic ferrites.

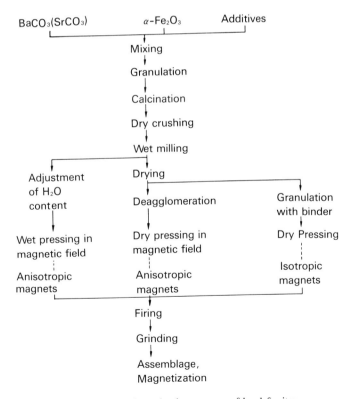

Fig. 15.8 General production process of hard ferrites.

important factors. Among others, hexagonal particles are preferred. The mechanical orientation techniques and that combined with a magnetic field are also studied extensively. As mentioned before, high fired density is important to attain high Br. The easiest way to achieve high fired density is to choose a high firing temperature. However, because high temperature sintering accelerates undesirable grain growth which lowers the coercive force, relatively, low temperatures ($\fallingdotseq 1{,}200°C$) aided by the additives are adopted in the practical manufacturing process. Additives for low temperature sintering are selected with reference to liquid phase sintering, and SiO_2 and $CaCO_3$ are found to be effective in yielding high-performance hard ferrites.[38,39,43]

15.3.2 Coercive Force

The intrinsic coercive force is a function of magnetocrystalline anisotropy and ratio of single domain particles, as shown by the following equation:

$$_iH_c = k \cdot n_s \cdot (2K/I_s) \tag{15.3}$$

where, $_iH_c$: intrinsic coercive force

n_s: contribution of single domain particles

K: magnetocrystalline anisotropy

I_s: saturation magnetization.

Magntocrystalline anisotropies of Sr ferrite and Ba ferrite are 3.7×10^6 erg/cm³ and 3.3×10^6 erg/cm³, respectively. Thus, the value of $2K/I_s$ for Sr ferrite is about 10 % greater than that for Ba ferrite; thus, Sr ferrite is expected to have higher $_iH_c$ than that of Ba farrite as shown in Fig. 15.9.

It is of prime importance to increase n_s in order to achieve high $_iH_c$. In other words, when each grain of fired polycrystalline body is composed of single domains, the highest

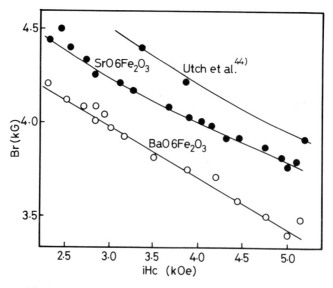

Fig. 15.9 Relationship between $_iH_c$ and Br of hard ferrites.

$_iH_c$, calculated to be about 8,100 Oe in the case of Sr ferrite, is expected.

In hard ferrite materials, critical size for single domain is thought to be about 1 μm. Under these situations control of powder properties, which include size and its distribution, shape, capability to orient, and compressibility, are most important.

It should be noted that high performance Sr-ferrite magnets are being manufactured using classified milled-powder.

In the present manufacturing process, calcination temperatures are usually higher than sintering temperatures in order to finish the ferritisation completely during calcination. Thus, the comminution process is important to control the powder characteristics which strongly influence the magnetic properties of the final products. Furthermore, in order to attain high $_iH_c$, the effects of some additives have been extensively studied, and the addition of SiO_2, Al_2O_3 and Cr_2O_3 is found to be effective.[45,46] These additives may behave as grain growth inhibitors, or may introduce higher K values than the non-doped materials. Recently, hard ferrite powders, which are obtained by glass-ceramics techniques,[47] coprecipitation[48] and sol-gel[49] methods, have been developed for the perpendicular magnetic recording media.

15.4 Concluding Remarks

The demand for both soft and hard ferrites has been growing and ferrites will expand markedly in both quantity and extent of application as the need for ferrites of higher quality increases. Raw materials, of which iron oxide is a major constituent, play a decisive role in improving the quality as well, as lowering the cost of ferrites. By combinations of improved raw materials, compositional and processing improvements, new classes of both soft and hard ferrites will be developed. Furthermore, new applications of ferrites such as ferrite carriers for electrophotocopy and biochemical applications will be expanded.

Acknowledgements

The authors wish to express their gratitude to Prof. Takashi Yamaguchi of Keio University for his help in the preparation of the manuscript.

References

1) A. Globus : "Some physical considerations about the domain wall size theory of magnetization mechanisms," J. de Phys., **38** (1977) C1-1.
2) A. D. Giles and F. F. Westendorp : "Simultaneous substitution of cabalt and titanium in linear manganese zinc ferrites," J. de Phys., **38** (1977) C1-47.
3) T. G. W. Stijntjes, J. Klerk and A. B. van Groenou : "Permeability and conductivity of Ti-substituted MnZn ferrites," Philips Res. Rep., **25** (1970) 95.
4) D. Stoppels and P. G. T. Boonen : "The influence of the second-order magnetocrystalline anisotropy on the initial magnetic permeability of MnZn ferrous ferrite," J. Magn. Magn. Mater., **19** (1980) 409.
5) K. Ohta : "Magnetocrystalline anisotropy and magnetic permeability of MnZnFe ferrites," J. Phys. Soc. Jpn., **18** (1963) 685.
6) K. Aso : "Mechanically induced anisotropy and its effect on magnetic permeability in

7) D. Stoppels, U. Enz and J. P. M. Damen : "Stress dependence of the magnetic permeability of MnZn ferrous ferrites," J. Magn. Magn. Mater., **20** (1980) 231.

8) E. G. Visser : "The stress dependence of the domain structure and the magnetic permeability of monocrystalline MnZnFe, II ferrite," J. Magn. Magn. Mater., **26** (1982) 303.

9) E. Röss : "Magnetic properties and microstructure of high permeability MnZn ferrites," FERRITES Proc. Int. Conf. Ferrites (1970) 187.

10) D. J. Perduijn and H. P. Peloschek : "MnZn ferrites with very high permeabilitiies," Proc. Brit. Ceram. Soc., **10** (1968) 2636.

11) A. Morita and A. Okamoto : "Effect of oxygen content on the properties of low loss MnZn ferrite," FERRITES Proc. Int. Conf. Ferrites (1980) 313.

12) P. K. Gallagher, E. M. Gyorgy and D. W. Johnson, Jr. : "Relation between magnetic permeability and stoichiometry of $Mn_{0.510}Zn_{0.417}Fe_{2.073}O_{4+\sigma}$," Am. Ceram. Soc. Bull., **57** (1978) 812.

13) T. Tanaka : "Sintering of MnZn ferrites with high density and predetermined oxygen contents," Funtai oyobi Funmatsuyakin, **25** (1978) 26.

14) E. Röss : "Eddy current and hysteresis losses in manganese zinc ferrites," FERRITES Proc. Int. Conf. Ferrties (1970) 187.

15) T. G. W. Stijnties and J. J. Roelofsma : "Low loss power ferrites for frequencies up to 500 kHz," FERRITES Proc. Int. Conf. Ferrites (1984) (in print).

16) T. Nomura, K. Okutani, T. Kitagawa and T. Ochiai : "Sintering of MnZn ferrites for power electronics," Am. Ceram. Soc. Fall Meeting (1982).

17) P. J. L. Reijnen : "Sintering behaviour and microstructures of sluminates and ferrites with spinel structure with regard to deviation from stoichiometry," Science of Ceramics, **4** (1968) 169.

18) P. E. C. Franken and N. van Doveren : "Determination of the grain boundary composition of soft ferrties by Auger electron spectroscopy," Ber. Dt. Keram. Ges., **55** (1978) 287.

19) M. Kakizaki, K. Ohya and K. Okutani : Kogyo Zairyo, **32** (1984) 52.

20) D. Stoppels, P. G. T. Boonen, J. P. M. Damen, L. A. H. van Hoof and K. Prijs : "Monocrystalline high-saturation magnetization ferrites for video recording head application," J. Magn. Magn. Mater., **37** (1983) 123.

21) U. König : "Eisen (III) oxid als Rohstoff für magnetische Werkstoffe," Ber. Dt. Keram. Ges., **55** (1978) 220.

22) T. Yamaguchi and T. Nomura : "Characterization and sintering of ferrite raw material powders," FERRITES Proc. Int. Conf. Ferrites (1980) 46.

23) U. Wagner : "Aspect of the correlation between raw material and ferrite properties, II," J. Magn. Magn. Mater., **23** (1981) 73.

24) B. B. Yu and A. Goldman : "Effect of processing parameters on morphology of MnZn ferrite particles produced by hydroxide-carbonate coprecipitation," FERRITES Proc. Int. Conf. Ferrites (1980) 68.

25) G. C. Jain, B. K. Das and N. C. Goel : "Effect of MoO_3 addition on the grain growth kinetics of a manganese zinc ferrite," J. Am. Ceram. Soc., **62** (1979) 79.

26) R. G. Wymer and J. H. Coobs : "Preparation, coating, evaluation and irradiation testing of sol-gel oxide microspheres," Proc. Brit. Ceram. Soc., **7** (1967) 61.

27) P. K. Gallagher, D. W. Johnson, Jr., E. M. Vogel and F. Schrey : "Microstructure of some freeze dried ferrites," Int. Mater. Symp., **6** (1977) 423.

28) U. Wagner : "Spray firing for preparation of presintered powder for soft ferrites," J. Magn. Magn. Mater., **19** (1980) 99.

29) M. J. Ruthner : "The importance of hydrochloric acid regeneration processes for the industrial production of ferric oxides and ferrite powders," FERRITES Proc. Int. Conf. Ferrites (1980) 64.

30) T. Ochiai and K. Okutani : "Ferrites for high frequency power supplies," FERRITES Proc. Int. Conf. Ferrites (1984) (in print).

31) M. Torii and R. Ishii : Denshi-Zairyo, **12**, 7 (1973) 95.

32) M. Torii, U. Kihara and I. Maeda : "New process to make huge spinel ferrite single

crystals," IEEE Trans. Magn., **MAG-15** (1979) 1873.

33) T. J. Berben, D. J. Perduijn and J. P. M. Damen : "Composition-controlled Bridgman growth of MnZn ferrite single crystals," FERRITES Proc. Int. Conf. Ferrites (1980) 722.

34) T. Kobayashi and K. Takagi : "Crystal growth of MnZn ferrite by the traveling solvent zone melting method," J. Crystal Growth, **62** (1983) 189.

35) S. Matsuzawa and K. Kozuka : "Method for producing ferrite single crystals by solid-solid reaction," FERRITES Proc. Int. Conf. Ferrites (1984) (in print).

36) L. J. Brady : "The constituents of $BaO5.5Fe_2O_3$ magnets," J. Mater. Sci., **8** (1973) 993.

37) J. S. Reed and R. M. Fulrath: "Characterization and sintering behavior of Ba and Sr ferrites," J. Am. Ceram. Soc., **56** (1973) 207.

38) F. Haberey and F. Kools : "The effect of silica addition in M-type ferrites," FERRITES Proc. Int. Conf. Ferrites (1980) 356.

39) F. K. Lotgering, P. H. G. M. Vromans and M. A. H. Huyberts : "Permanent-magnet material obtrained by sintering the hexagonal ferrite $W-BaFe_{18}O_{27}$," J. Appl. Phys., **51** (1980) 5913.

40) G. Albanese, M. Carbucicchio, F. Bolzoni, S. Rinaldi, G. Sloccari and E. Lucchini : "Magnetic properties of aluminium substituted Zn_2-W hexagonal ferrites," Physica BC, **86/88** (1977) 941.

41) S. Dey and R. Valenzuela : "Magnetic properties of substituted W and X hexaferrites," J. Appl. Phys., **55** Pt 2B (1984) 2340.

42) T. Besagni, A. Deriu, F. Licci, L. Pareti and S. Rinaldi : "Nickel and copper substitution in Zn_2-W," IEEE Trans. Magn., **MAG-17** (1981) 2636.

43) R. H. Arendt : "Liquid-phase sintering of magnetically isotropic and anisotropic compacts of $BaFe_{12}O_{19}$ and $SrFe_{12}O_{19}$," J. Appl. Phys., **44** (1973) 3300.

44) B. Utsch and R. Corbach : "New magnetic properties of ferrites," FERRITES Proc. Int. Conf. Ferrites (1984) (in print).

45) L. G. van Uitert and F. W. Swanekamp : "Permanent magnet oxides containing divalent metal ions, II," J. Appl. Phys., **28** (1957) 482.

46) K. Haneda and H. Kojima : "Intrinsic coercivity of substituted $BaFe_{12}O_{19}$," Jpn. J. Appl. Phys., **12** (1973) 355.

47) B. T. Shirk and W. R. Buessem : "Magnetic properties of barium ferrite formed by crystallization of a glass," J. Am. Ceram. Soc., **53** (1970) 192.

48) T. Takada and M. Kiyama : "Preparation of ferrites by wet method," FERRITES Proc. Int. Conf. Ferrites (1970) 69.

49) K. Oda, T. Yoshio and K. Takahashi : "Study on $Ba_{1-x}Sr_xFe_{12}O_{19}$ particles precipitated from a glass," Funtai oyobi Funmatsuyakin, **29** (1982) 39.

Authors' Profile

---- Editor ----

Shinroku Saito was born in Tochigi, Japan, on March 30, 1919. He received the B.S. degree in 1943 from Tohoku University, Sendai, Japan. Following his graduation from Department of Technology, Tohoku University, he worked for two years as a research associate in the Central Research Laboratory for Aeronautics attached to the Cabinet. In 1945, he was appointed to an associate professor of a private college, the Technical College of Kugayama. In 1950, his 30 years' career in Tokyo Institute of Technology, a national organization, started, where he also received Dr. Eng. degree in 1960. He joined there as a research associate and was appointed as an associate professor, then a professor, in Research Laboratory for Engineering Materials. By 1973, he was the director of the laboratory and in 1977, he was appointed as the president of Tokyo Institute of Technology and stayed there untill 1981. Since 1983 till present date, he has been working as the president of the Technical University of Nagaoka, Niigata, Japan.

His other current involvements are; in the private sector — Member of Technological Planning Committee (Technova Inc.), Chaiaman of Committee for Space Station Utilization (Mitsubishi Research Institute Inc.), in the public sector—Chaiman of Selection Committee of Garment and Apparel Culture Prize (Foundation Advanced Garment and Apparel Research (FAGAR)), President of Japan Fine Ceramic Association, President The Ceramic Society of Japan, Vice President of Fine Ceramics Center of Japan, Director of The Society of Non-Traditional Technology, Director of Industrial Research Institute, Advisor of Advanced Machining Technology & Development Association, Director of The Material Science Society of Japan, Director of Japan Society for Science Studied Policy and Research Management, and in the governmental & quasigovernmental sectors—Member of Council of University and College Establisment/Member of Committee for Postgraduate Course (Ministry of Education and Culture), Member of Industrial Technology Council (Ministry of International Trade and Industry), Councilor, Member of Council for Aeronautics, Electronics and other Advanced Technology/Councilor of Japan Space Utilization Promotion Center/Councilor of Space Activities Commission/Councilor of Resources Council (Science and Technology Agency), Member of Committee of Inspection and Coordination & Council for Science & Technology (Prime Minister's Office), Councilor, Chairman of Committee for Space Shuttle Utilization (National Space Development Agency), Councilor of Japan Key Technology Center, Chairman of Committee for Technology Development of the Small Business Corporation, Member of Committee of Creative Science of Research Development Corporation of Japan, Vice Chairman of Advisory Committee/Member of 136th Committee on Future Oriented Machining/Member of 149th Committee on High Technology & International Issues (Japan Society for the Promotion of Science), Member of Committee for Selection of Members (Science Council of Japan), Councilor (Japan International Cooperation Agency), Member of 1st Committee (Association of National Universities).

---- Introduction ----

Kiyoshi Okazaki is Vice-President and Professor of Electronic Ceramics at The National Defense Academy, Yokosuka, Japan. He earned his B. S. degree in electrical engineering in 1949 and his Dr. Eng. degree in 1959, both from Kyoto University, Japan. Dr. Okazaki was with Murata Mfg. Co. before joining The National Defense Academy in 1955. He was named Associate Professor in 1958, Professor in 1963, Dean of the Graduate School of Scientific and Engineering Course Programs at The National Defense Academy in 1983, and to his present position in 1984. He has served as the Chairman of Dept. of Electrical Engineering during 1963-65. He has been a Fellow of the American Ceramic Society since 1980.

AUTHORS' PROFILE

―――― 1 ――――

Hisao Banno is Director, R&D Nagoya Labs, NTK Technical Ceramics Division, NGK Spark Plug Co., Ltd., Nagoya, Japan. He earned his B.S. degree in electrical engineering from Nagoya University, Nagoya, Japan, and joined NGK Spark Plug Co., Ltd., in 1957. His work has been mainly on piezoelectric ceramics and composites.

He received the 3rd R. M. Fulrath Award in recognition of his study on "Development of Modified Pb (Zr, Ti) O_3 ceramics" from the University of California at Berkeley, U.S.A., and the American Ceramic Society in 1980.

―――― 2 ――――

Noboru Ichinose is a professor in the School of Science and Engineering at Waseda University, Tokyo, Japan. As a graduate of Waseda University, he earned the B.S. degree in applied physics in 1959 and the Ph. D. degree in physics in 1967.

He was with Toshiba Corp. prior to joining the university in 1985. His work concerns electronic ceramics such as ferrites, PTC and NTC thermistors, ceramic nonlinear resistors, piezoelectric and pyroelectric ceramic materials, and ceramic sensors. He received the 1st R.M. Fulrath Award in 1978.

―――― 3 ――――

Fumikazu Kanamaru is a professor of nonstoichiometric crystalline materials division, The Institute of Scientific and Industrial Research (ISIR), Osaka University, Osaka, Japan. He received the M.S. degree in physical chemistry and the Dr. Sci. degree from Osaka University, in 1958 and 1963, respectively.

Prior to joining ISIR in 1980, he was an assistant professor of synthetic inorganic materials division, ISIR, in 1966 to 1976, and a professor in noncrystalline materials Laboratory, Okayama University, Okayama, Japan, in 1976 to 1980.

He received the 1st R.M. Fulrath Award in 1978.

―――― 4 ――――

Sadayuki Takahashi was born in Osaka, Japan, on May 6, 1941. He received the B.S. degree in physics from Osaka University, Osaka, Japan, in 1964.

He joined NEC Corporation, Japan, in 1964, and has engaged in research and development on piezoelectric ceramic materials, thin films and their application devices. He is presently a research specialist at the Fundamental Research Laboratories.

Mr. Takahashi is a member of the Institute of Electronics and Communication Engineers of Japan, Japan Society of Applied Physics, the Acoustical Society of Japan and the Japan Society of Powder and Powder Metallurgy.

He received the 8th R.M. Fulrath Award in 1985.

―――― 5 ――――

Katherin T. Faber received her Ph. D. degree in materials science at the University of California, Berkeley, U.S.A., in 1982. Dr. Faber joined the faculty of the Department of Ceramic Engineering at The Ohio State University, Columbus, U.S.A. She has also held positions at The Carborundum Company and Lawrence Livermore Laboratory. Her interests include toughening mechanisms in brittle materials, thermal stress and reliability, and the mechanical behavior of electronic materials.

―――― 6 ――――

Robert F. Davis is Professor of Materials Engineering and Director, Materials Research Center, North Carolina State University, U.S.A. He received his Ph. D. degree from the University of California, Berkeley, U.S.A.

His principal research interests are deformation mechanisms in ceramics, ceramic thin films and coatings and solid state diffusion.

Dr. Davis is a fellow of the American Ceramic Society.

AUTHORS' PROFILE

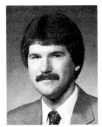

Calvin H. Carter, Jr. is visiting Research Professor of Materials Engineering at North Carolina State University, U.S.A., where he received his Ph. D. degree.

His principal research interests are kinetics and mechanisms of deformation in nonoxide ceramics, growth and characterization of semiconductor thin films and high resolution TEM.

———————— 7 ————————

James F. Shackelford is a Professor in the Division of Materials Science and Engineering and the Associate Dean for Undergraduate Studies in the College of Engineering at the Davis Campus of the University of California, U.S.A. He received his B.S. and M.S. degrees in ceramic engineering at the University of Washington, U.S.A., in 1966 and 1967, respectively, and a Ph. D. degree in materials science and engineering at the Berkeley Campus of the University of California in 1971.

During 1972 and 1973, he was a postdoctoral fellow at McMaster University in Hamilton, Ontario, in Canada. Since 1973, he has been associated with the University of California at Davis as Assistant Professor (1973-1979) and Associate Professor (1979-1984) and assumed the positions of Professor and Associate Dean in 1984. Professor Shackelford's research interests include glass, biomaterials, and nondestructive testing.

He is the author of over 40 publications in the field of materials including the textbook *Introduction to Materials Science for Engineers*.

———————— 8 ————————

Donald M. Smyth is Director of the Materials Research Center and Professor of Chemistry and of Metallurgy and Materials Engineering at Lehigh University in Bethlehem, Pennsylvania, U.S.A. He received the B.S. degree in chemistry with highest distinction from the University of Maine, U.S.A., in 1951, and the Ph. D. degree in inorganic chemistry from M.I.T., U.S.A., in 1954. From 1954 to 1971, he was at the Research and Development Laboratories of the Sprague Electric Company in North Adams, Massachusetts, U.S.A. His research work there involved the solid state chemistry of solid ionic conductors and of capacitor dielectric materials such as Ta_2O_5 and $BaTiO_3$. He joined the faculty at Lehigh University in 1971 and has continued to study the defect chemistry and electrical properties of transition metal oxides.

Dr. Smyth is a Life Fellow of the American Institute of Chemists, and is a member of the American Ceramic Society, the American Chemical Society, and the Electrochemical Society.

———————— 9 ————————

Yoshiharu Ozaki is a member of the Engineering faculty of the Seikei University, Musashino, Tokyo, Japan. He attended the Tokyo Institute of Technology, Tokyo, Japan, and he received the B.S. degree in 1966, the M.S. degree in 1968 and the Dr. Eng. degree in 1971, all in ceramic engineering. After two years in a post doctoral fellow at Research Laboratory of Engineering Materials, Tokyo Institute of Technology, he joined staff of the Seikei University as associate professor of ceramic engineering laboratory in the Department of Industrial Chemistry. In 1984, he was promoted to professor. During 1974-1975, he was visiting scientist at Centre National de La Recherche Scientifique, Grenoble, France. He has been interested in the development of the ceramic processing technology, and worked in the fine powder preparation from organometallic compound, metal alkoxides since 1975. Above these research activities, he has served for various professional societies. For the Ceramic Society of Japan, he has served as all officers including chairman of Bulletin during 1984-1985. And he has served as a member of Industry-University Cooperative Committees of Japan Society for the Promotion of Science since 1975, and a director of La Societe Franco-Japonaise des Techniques Industrielles during 1982-1984.

———————— 10 ————————

Tamotsu Ueyama received his B.S. degree from Tokyo Science University, Japan. From 1954 to 1960, he worked as research staff of Hitachi Central Research Laboratory, Hitachi, Ltd., Japan, for the research on Hitachi Solution Properties of high Polymer. From 1961 to 1977, he worked for Research and Development of Industrial Laminates & Printed Circuits Board in Shimodate Works, Hitachi Chemical Co., Ltd., Japan.

He is a Senior Researcher of Ibaraki Research Laboratory, Hitachi Chemical Co., Ltd. for Research and Development of alumina ceramic and thick film substrate.

Mr. Ueyama is a member of the Society of Ceramics of Japan and the Polymer Science Japan.

Hiroshi Wada received the B.S. degree in ceramic science from Tokyo Institute of Technology, Tokyo, Japan, in 1968.

He is a researcher of processing technology for ceramics at Ibaraki Research Laboratory, Hitachi Chemical Co., Ltd., Japan. He worked for the research and development of alumina ceramics from 1968 to 1976, PZT ceramics from 1976 to 1977, and electronic alumina ceramics from 1977 to 1983. Since 1983, he has worked for the research and development of zirconia ceramics.

Masahiko Shimada, born in 1940 at Obama, Fukui, Japan, is a professor of ceramic Sciences of the Faculty of Engineering, Tohoku University, Sendai, Japan. He graduated from the Geological and Mineralogical Institute, Kyoto University, Kyoto, Japan, in 1964 and received his M. Sc. degree there in 1966. In 1971, he received his Dr. Sci. degree from Kyoto University. Prior to joining Tohoku University, in 1983, he was with Osaka University, Osaka, Japan, during 1966-1982. During 1976-1977, he was visiting scientist at Pennsylvania State University, U.S.A. His research includes high pressure synthesis, hot isostatic pressing, and solid state bonding between ceramic and metal.

He received the 8th R.M. Fulrath Award in 1985.

Katsuaki Suganuma is a research associate of the Research Center of High Pressure Synthesis of the Institute of Scientific and Industrial Research, Osaka University, Osaka, Japan. He graduated from the Faculty of Engineering, Tohoku University, Sendai, Japan, in 1977 and received his Dr. Eng. degree from Tohoku University in 1982.

Taira Okamoto is a professor of Metallic Materials and a director of the Institute of Scientific and Industrial Research, Osaka University, Osaka, Japan. He graduated from the Faculty of Engineering, Osaka University in 1951 and received his Dr. Eng. degree from Osaka University in 1962.

Mitsue Koizumi is a professor of ceramic science of the Institute of Scientific and Industrial Research, Osaka University, Osaka, Japan. He graduated from the Geological and Mineralogical Institute, Kyoto University, Kyoto, Japan, in 1945 and received his Dr. Sci. degree from Kyoto University in 1958. From 1956 to 1958, he was a visiting professor at Pennsylvania State University, U.S.A.

Takaki Masaki is a research associate at the Technical Development Department of Toray Industries, Inc., Japan.

He received his B.S. degree in material science in 1967, his M.S. degree in material science in 1969, from Osaka University, Osaka, Japan.

He joined Toray Industries, Inc. in 1969. His research activities include powder processing and fabrication, sintering and microstructure development, in ZrO_2 ceramics. His current interest is to develop high toughened ZrO_2 ceramics.

Keisuke Kobayashi is an assistant general manager at the Technical Development of Toray Industries, Inc., Japan.

He received his B.S. degree in metallurgy in 1955, his Dr. Eng. degree in metallurgy in 1961, from the University of Tokyo, Tokyo, Japan.

He joined Toray Industries, Inc. in 1962. His research activities include the solid electrolyte of ZrO_2-Y_2O_3 system, toughened zirconia ceramics and its applications. His current interest is to develop ceramic powder materials and its characterization.

13

Shigetaka Wada is a senior researcher as well as the manager of the ceramics division of Toyota Central Research and Development Laboratories, Inc. (CRDL), Japan. He received the B. Sc. degree in physics from Nagoya University, Nagoya, Japan, in 1961. From 1961 to 1982, he worked at NGK Insulators Ltd., Japan, then joined Toyota CRDL in 1982.

He has been engaged in the research and development of hot pressing and hydrothermal syntheses of Al_2O_3, glass ceramics, semi-conductive barium titanate ceramics and its applications as the honeycomb structured heater, and ZnO varistor, as well as ZrO_2 ceramics for heating elements and oxygen sensors. His recent works subsequently emphasize the engineering ceramics processing of the materials and its applications for gas turbine engine.

14

Koichiro Takahashi is senior research associate of Muki-Zaishitsu Kenkyu-Sho (National Institute for Research in Inorganic Materials) at Tsukuba Academic Region, Japan. His expertise is in glass and ceramic science and ferroelectrics. He graduated from the Department of Applied Chemistry, School of Science and Technology, Waseda University, Tokyo, Japan, in 1963, then entered the doctorate program at the Tokyo Institute of Technology, Tokyo, Japan, in the same year. He received his Ph. D. degree in 1968 from Tokyo Institute of Technology in glass science. From 1973 to 1974, he served as visiting researcher at Materials Research Laboratory at the Pennsylvania State University, U.S.A., working with Profs. Robert E. Newnham and Leslie E. Cross.

15

Takeshi Nomura was born in Hokkaido, Japan, on March 8, 1952. He received the B.S. degree in 1975, the M.S. degree in 1977 and the Ph. D. degree in 1980, all from Keio University, Japan. Since 1980, he has been engaged in research and development of high-performance soft ferrites in the Ferrite Research and Development Laboratory, TDK Corporation, Japan.

Katsunobu Okutani was born in Hiroshima, Japan, on January 29, 1941. He received the B.S. degree from Shimane University, Japan, in 1963. He has worked for TDK Corporation, Japan, since 1963. He is now a chief engineer in the Ferrite Research and Development Laboratory, TDK Corporation. He is engaged in the development of soft ferrites.

Tatsushiro Ochiai was born in Hiroshima, Japan, on November 25, 1932. He received the B.S. degree in 1956 from Chiba University, Chiba, Japan. He has worked for TDK Corporation, Japan, since 1958. He is now a manager of the Ferrite Research and Development Laboratory. He is engaged in the development of soft ferrites.

Keywords Index

actuator 62
alumina powder 184
alumina substrate 184
applications 8
automotive components 227

casting 184
ceramic 62, 147, 227
ceramic materials 27
ceramic-metal joint 201
ceramic scissors 210
charge transfer type intercalation compound 41
composites 8
conductivity 147
creep 95

defects 147
deflection 76
deformation mechanisms 95
dispersion 184

electromechanical coupling 62
electron microscopy 95

feasibility/reliability/cost 227
ferrite 254
ferroelectrics 240
field induced strain 62
film 240
fine powder 165

gas sensor 27
glass 126

hexagonal 254
HIPing 201
humidity sensor 27

infrared rad sensor 27
ionic conductivity 147

lead germanate 240

material 8
metal alkoxides 165
microcracking 76
microstructure 210
multilayer 62
multilayer capacitor 165

non-crystalline 126

organic derivative 41
organic intercalation compound 41

partially stabilized zirconia 210
permeability 254
perovskite compound $BaTiO_3$ 165
piezoelectric 8
piezoelectricity 62
PLZT light switch 165
powder processing 210
properties 126
pyroelectrics 240

R-curve 76

semiconducting oxides 147

277

shrinkage 184
silicon carbide 95
silicon nitride 95
single crystal 254
solid-state bonding 201
speaker diaphram 165
spinel 254
stack 62
stereospecific polymerization 27
stress induced transformation 210
structure 126
surface roughness 184
tape-casting 62

temperature sensor 27
tetragonal zirconia polycrystal 210
theory 8
toughened ceramics 210
toughness 76
transformation 76

valence compensated perovskite 165

$YAlO_3$ 165

zirconia 76, 210

STAFFORDSHIRE
COUNTY REFERENCE
LIBRARY
HANLEY
STOKE-ON-TRENT